TRANSPORTER BRIDGES

TRANSPORTER BRIDGES

AN ILLUSTRATED HISTORY

JOHN HANNAVY

PEN & SWORD
TRANSPORT

AN IMPRINT OF PEN & SWORD BOOKS LTD.
YORKSHIRE – PHILADELPHIA

First published in Great Britain in 2020 by
Pen & Sword Transport
An imprint of
Pen & Sword Books Ltd
Yorkshire - Philadelphia

ISBN 978 1 52676 038 8

Typeset in Palatino 11/14
by Aura Technology and Software Services, India
Printed and bound in China through Printworks Global Ltd.

Pen & Sword Books Ltd incorporates the Imprints of Pen &
Sword Books Archaeology, Atlas, Aviation, Battleground,
Discovery, Family History, History, Maritime, Military, Naval,
Politics, Railways, Select, Transport, True Crime, Fiction,
Frontline Books, Leo Cooper, Praetorian Press, Seaforth
Publishing, Wharncliffe and White Owl.

For a complete list of Pen & Sword titles please contact

PEN & SWORD BOOKS LIMITED
47 Church Street, Barnsley, South Yorkshire, S70 2AS, England
E-mail: enquiries@pen-and-sword.co.uk
Website: www.pen-and-sword.co.uk

or

PEN & SWORD BOOKS
1950 Lawrence Rd, Havertown, PA 19083, USA
E-mail: Uspen-and-sword@casematepublishers.com
Website: www.penandswordbooks.com

Cover image: Looking along the cradled main beam of the
Puente Viscaya as the gondola docks at Gexto, (Las Arenas).
The bridge – the world's first transporter bridge – crosses
the River Nervión from Portugalete to Gexto a few miles
downstream from Bilbao in Northern Spain. It was opened in
1893 and still operates 24 hours a day, 7 days a week.

Title page image: A view of the docks at Nantes with the Pont
Transbordeur in the distance. This tinted postcard dates from
around 1952, by which time the bridge was approaching its
golden jubilee. It was dismantled in 1958, leaving only the
stone piers.

Contents page image: A W.D. & H.O. Wills cigarette card
from c.1915 showing a stylised interpretation of Ferdinand
Arnodin's Newport Transporter Bridge – No.30 in a series
showing engineering marvels.

Photograph & Illustration credits
All modern photographs and all historic images are,
unless otherwise stated, © John Hannavy Image Library,
or come from private collections. Other illustrations are
acknowledged as follows: Christophe Accart/ARTCAD 227;
www.alwayshobbies.com 10; Biblioteca Digital Luso-Brasileira
26 *bottom*; © Dr. Ron Callender 48 top 245; Catalyst Science
Discovery Centre 114, 115 *bottom*, 117 *top*, 119 *top*, 120
margins, 125 *top*, 126, 127 *top*, 128, 135, 139; Courtesy of
Tony Cook, Railway Signalman 129; © Lars Curfs 48 *bottom*;
© Daniel Darwall 43, 44; Dublin Port Archives 55, 262;
Durham Record Office 101, 254, 258; © Chris Gascoigne 60;
Ghent University Library BIB.037B001 50 *bottom*, 260;
© Margaret Ingham/Friends of Warrington Transporter Bridge
42 *bottom*; © Petra Klawikowski under Creative Commons
license 240; Lifschutz Davidson Sandilands 60, 263; Maison
du Transbordeur, Rochefort-sur-Mer 219 *bottom*, 226 *top*;
Newport Museum and Heritage Service 159 *top left*;
© National Trust Images, Edward Chambré Hardman
Collection 6; Ordnance Survey Crown Copyright © 1905 &
1926 reproduced with permission 196; Pixabay 243; Paul
Poirier 61 *middle & bottom*, 63; Puente Viscaya 71 *bottom*,
74, 75; Runcorn Historical Society 134 *top*; Staffordshire
Record Office 186, 187, 188, 192; © Alan Ratcliffe/Friends
of Warrington Transporter Bridge 211; © David A. Simm 40;
Teesside Archives, 162, 165, 167; *The News*, Portsmouth 52, 256;
Reproduced with kind permission of Unilever from originals
in Unilever Archives 194, 198, 199, 203; Kris Ward 202 *top*;
© Waterways Images 191; Wikipedia Commons 56, 249.

CONTENTS

E. Chambré Hardman

PREFACE

Amongst the many things I never anticipated doing in my life was standing on a temporary platform high above the Charente river in Western France photographing workmen far below me as they restored the 117-year-old Rochefort-sur-Mer transporter bridge.

One of the more unusual examples of Victorian and Edwardian innovation, transporter bridges were seen as a simple and innovative solution to a very big problem.

452 ROUEN. — Le Quai du Havre et le Transbordeur. — LL.

Opposite: Edward Chambré Hardman's dramatic study of the Widnes–Runcorn Transporter Bridge in the late 1950s. (© National Trust Images, Edward Chambré Hardman Collection)

Left and below: Two early postcard views of the Rouen Transbordeur by French postcard publishers Léon & Lévy.

35 ROUEN. — Ensemble du Pont Transbordeur. — LL.

For just a few years, between 1893 and 1916, a number of these massive structures were constructed across Europe, with a single 'bridge ferry' built in the United States, and four in South America.

They were, however, quickly overtaken by the rapid increase in road traffic in the early twentieth century, rendering them slow and uneconomic, and today only a few rare examples survive.

During their working lives, however, these bridges were the subjects of hundreds of postcards which today provide us with a fascinating visual history of an ingenious solution to a simple problem – how to bridge docks and rivers in such a way that the passage of tall-masted ships in and out of busy ports would not be impeded.

For the few decades during which they operated, these huge structures were not just popular means of transportation, they were objects of considerable fascination and very quickly became popular and challenging subjects for photography. The few surviving bridges still hold that fascination for photographers today.

Opposite: Warrington's Bank Quay Transporter Bridge was last used in the 1960s. Rusting and deteriorating, the Grade II* listed bridge is now on the 'at risk' register of historic structures.

Below: Meccano construction kits – the ultimate boys' toys of their age – could be used to built all sorts of mechanical objects. Between 1918 and 1953 the company published several sets of plans for different designs of transporter bridges. The model of the Widnes–Runcorn Bridge *(top)* featured in their catalogues from 1928, ten years after the plans for the model of the Rouen bridge *(bottom left)* first appeared. The splendid model *(bottom right)*, which also first appeared in 1928, appears to be loosely based on the Ponte Alexandrino in Rio de Janeiro. The Newport bridge was also the subject of a feature article in *Meccano Magazine* in June 1959.

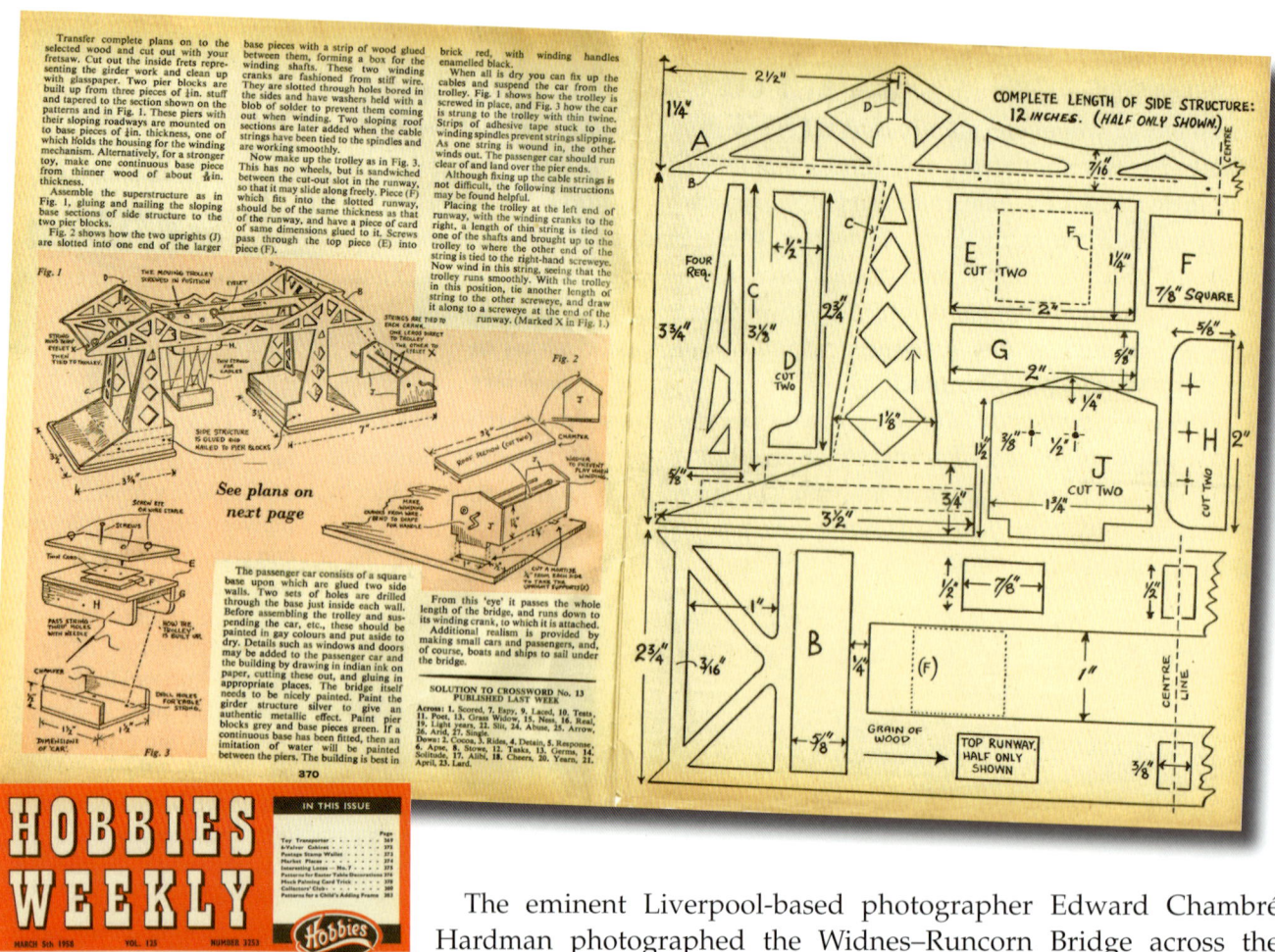

Instructions for making a simple wooden transporter bridge 'inspired by the one in Middlesbrough' – and needing nothing more than 1/8in plywood and a fretsaw – was featured in the 5 March 1958 issue of *Hobbies Weekly*.

The eminent Liverpool-based photographer Edward Chambré Hardman photographed the Widnes–Runcorn Bridge across the Mersey, one of a series of studies he made in the late 1950s, most probably in 1959. It was taken from the approach ramp on the Runcorn side, with the heavily industrialised skyline of Widnes visible through the framework. Just a few years later, the magnificent bridge was dismantled, replaced by a road bridge.

The Hungarian-born constructivist photographer László Maholy-Nagy was one of many others who saw their creative potential. He travelled to Marseille to produce an important series of images. From 1923 until 1928 he taught at the hugely influential Bauhaus school of art and design in Berlin – he was fascinated by technology and used his work to create a fusion of ideas where art and industry met. Another Bauhaus photographer, the Austrian-born Herbert Bayer, shared Maholy-Nagy's fascination for the Marseille bridge, photographing it in 1928.

The era of the transporter bridge was so brief, and the numbers planned and built so few, that there was not even time for a descriptive lexicon to be compiled and agreed upon. In addition to the term 'transporter bridge' – 'transbordeur' in French and 'transbordador' in Spanish – they were variously known as 'aerial ferries', 'ferry bridges', 'aerial cars', 'flying bridges', 'flying ferries', 'suspended ferries', 'transporting bridges', 'hovering ferries', 'hanging

38 ROUEN. — *Le Pont Transbordeur.* — LL.

In the days before health and safety considerations were even thought of, a single rope was all that protected passengers and vehicles travelling across the Seine on the Pont Transbordeur at Rouen. This is one of an extensive series of views produced by leading French postcard publishers Léon & Lévy shortly after the bridge opened.

23 BREST. — *Le Transbordeur dans l'Arsenal.* — The Trans-shipping in the Arsenal. — LL.

The Pont à Transbordeur at Brest gave access to the French Naval Arsenal. This postcard was published c.1910, not long after the bridge was opened, having been re-located from Bizerte in Tunisia and re-erected in the dockyard.

bridges', 'conveyor bridges', 'traveller suspension bridges' and probably some others as well.

The hanging structure which carried people and vehicles was variously referred to as the 'cage', 'moving platform', 'gondola', 'nacelle', 'travelling car', 'suspension car', 'transporting basket' and any one of a number of other vaguely descriptive terms.

From the opening of the very first transporter bridge in 1893, interest in them was considerable, both from a construction point of view – they appeared to offer a low-cost means of crossing the world's many navigable rivers – and from a commercial viewpoint, they seemed likely to recoup their construction and operating costs relatively quickly. Plans to build at least 35 transporters were initiated worldwide, but only 21 were ever completed. Details of all of them can be found in the Fact Sheets at the conclusion of this volume.

Perhaps because there were so few of them, public fascination with these strange structures was quickly recognised and tourists bought postcards of them in their millions. Less expected was that the fascination would last longer than many of the bridges themselves, with boys of all ages well into the 1960s wishing they had the Meccano 'Super Kits' which would enable them to construct their own models of these structures. The kit to build the Widnes–Runcorn Bridge, for example, could still be purchased even after the bridge itself had been demolished. For those of more modest means, there was also the challenge of building less sophisticated wooden models of them, following instructions featured in magazines such as *Hobbies Weekly*.

Unpicking the story of the transporter bridge has been a journey of discovery, and gathering together a representative collection of images has been a fascinating challenge. That the brief 'golden age' of the transporter bridge coincided with the golden age of the Edwardian postcard is fortuitous, leaving us with many splendid images of these unusual bridges.

The sheer numbers of postcards, souvenirs and other mementos which were produced is a testament to the fascination which these great structures held for our Edwardian counterparts – those same people who were fascinated by steam railways and by canals. and who bought postcards of them in their hundreds of millions. These bridges also enjoyed their heydays before passing out of fashion.

Steam on the railways gave way to diesel and electric traction, canal transport gave way to rail and road transport, and postcards were eclipsed, to an extent, by the democratisation of photography, which came about with the introduction of the Kodak box camera and its successors.

But things have a habit of turning full circle. We now visit preserved steam railways in our hundreds of thousands each year, with lines being re-opened and new locomotives being built to cater for an expanding leisure market. Canal restoration projects have led to new innovative solutions to the centuries' old challenges of moving water over hills – the spectacular Falkirk Wheel, already a hugely popular twenty-first century tourist destination, being a significant example.

Could the time also be ripe for the rediscovery of the transporter bridge as a tourist attraction? The next few years could be very interesting in that respect. Certainly, the experiences of travelling across a river just a few feet above the surface of the water, or climbing the towers and walking across the suspension beam are both rare and exhilarating.

A case is currently being made for World Heritage Site recognition to be given to all the surviving transporters – so far only the very first one, the Puente Viscaya, or Bizkaia, at Portugalete near Bilbao in Spain, is listed by UNESCO – and it is hoped that this attempt will be successful in time, and that these unique structures will get the sort of protection they so clearly require before it is too late. A campaign group, Friends of the Warrington Transporter Bridge, is working hard to raise the profile of the British transporter bridge most at risk. We wish them every possible success in their endeavours before their magnificent structure deteriorates beyond salvation.

A UNESCO inscription for the world's surviving transporters, if ever granted, would be one of the most geographically wide-ranging, covering sites in the UK, France, Spain, Germany and Argentina. However, the process towards achieving World Heritage Site status is a long and difficult one, especially for citations which would cross national borders.

The first stage is getting support from each country for the inclusion of their sites on the 'Tentative List'. So far France, Germany, Argentina and the UK have not got that far.

Argentina's only surviving transporter bridge is the Puente Nicolás Avellaneda in Buenos Aires. This postcard was published in the 1920s. The bridge is currently nearing the end of a major restoration programme aimed at returning it to service as a tourist attraction in 2018 – part of the regeneration programme for the city's docklands.

BUENOS AIRES (Rep. Argentina)
Puente Avellaneda

Britain's last citation for an industrial monument was the Forth Railway Bridge in 2015. It had been on the Tentative List for a surprisingly short space of time, while multi-site industrial groupings such as the important remains of the Welsh Slate Industry have just been selected as the UK government's 2018 nomination after years of campaigning.

As has been the case with so many of my projects over the years, the initial research that underpins this book was undertaken because I could not find convincing answers available in print to my questions about the origins and history of transporter bridges while I was engaged on an earlier project.

What I had not anticipated was the number of proposed transporter bridges which never got beyond the drawing board. Their inclusion in this book enriches the story immeasurably.

When I tried to follow up leads and check information back to primary sources, I discovered that a surprising amount of what has been published over the years was vague, unrevealing and, sometimes, simply wrong.

One of the bugbears of the age in which we live is that once something has been published, accurate or otherwise – especially on-line – it eventually gets repeated often enough for it to assume the mantle of apparent authenticity.

Unravelling fact from fiction has been one of this project's greatest challenges, and the story which emerges is a fascinating one. Hopefully, in answering my own questions it will turn out that, on the pages which follow, I have answered some of yours as well.

I also hope that publishing this book might help bring further information to light so that a much more complete history of the world's transporter bridges might one day be written.

I could not have completed this project without a great deal of help from a great many people and organisations. I am grateful to Christophe Accart/Artcad-etudes, William Alschuler, Eduardo Alvelo, Harry Arnold, William

A souvenir postcard of the Widnes–Runcorn transporter bridge across the Mersey, which was opened in 1905. This composite card was photographed and marketed by W. Hall of Widnes shortly after the bridge entered service.

Bill/Glasgow University Archives, Ross Bullock/Runcorn & District History Society, Turtle Bunbury, Dr Ron Callender, John Carruthers, John Clayson/ Tyne & Wear Archives & Museums, Tony Cook, Daniel Darwall, Dorset History Centre, Liz Bregazzi and staff/Durham Record Office, Femke Van der Fraenen/Ghent University Library, Chris Gascoigne, Jennifer Glanville/ University of Reading Museums and Special Collections Service, Javier Goitia/Consultant Engineer Puente Viscaya, David Hando/Friends of Newport Transporter Bridge, Alan Hayward and Margaret Ingham/Friends of Warrington Transporter Bridge, John Ivison and his team/Tees Transporter Bridge, Lar Joye/Dublin Port Company, Madelaine Lecouturier/Unilever Archives, Mike Lewis/Culture and Continuing Learning Manager Newport Council, Lifschutz Davidson Sandilands, Paul Meara/Catalyst Science Discovery Centre Widnes, Andrew Meek/www.alwayshobbies.com, Chris Mills/Ribble Steam Railway, Robert Morris/National Tramways Museum, Tania Parker/National Railway Museum, National Trust Images, Professor Roland Paxton, Paul Poirier, Pauline Prion/OPPIC, Craig Sherwood, Staffordshire County Archives, Kimberley Starkie/Teesside Archives, Andrew Tweedie/Gracesguide.co.uk, Kris Ward/Leedsengine.info, Tosh Warwick/ Heritage Development Officer Middlesbrough Council, Jonanthan Wilson/ Black Country Living Museum, John Vignoles, Mark Waldron/Evening News Portsmouth, and everyone else who has added snippets of information, supplied me with revealing documents, alerted me to images, corrected my misunderstandings, allowed me access to their archives and guided me through the engineering challenges with which the bridge-builders had to contend.

And very special thanks as always, to my wife, Kath, for her constant support and encouragement, and to Karen Easton for all her help with translations.

John Hannavy, Great Cheverell, 2019

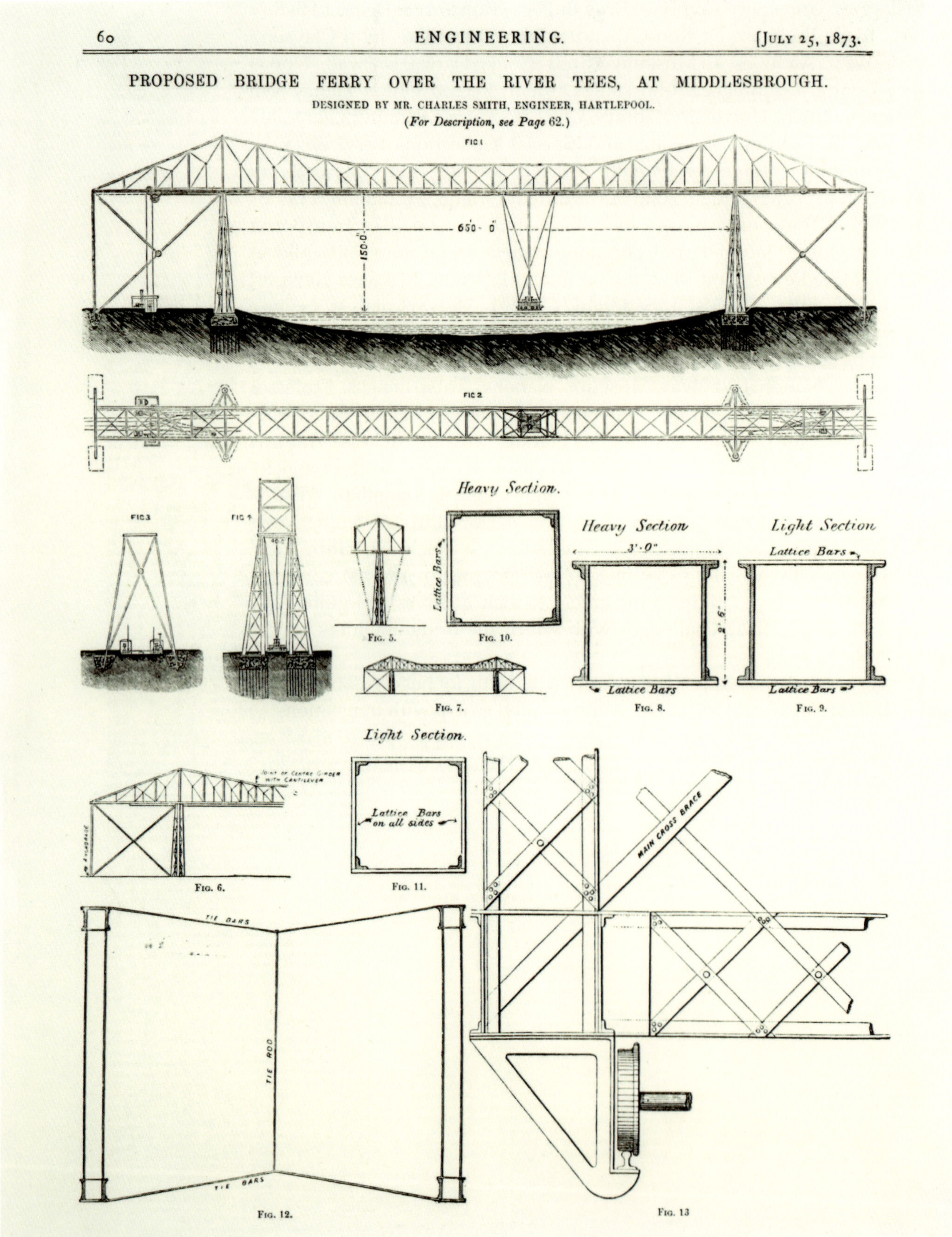

60   ENGINEERING.   [JULY 25, 1873.

PROPOSED BRIDGE FERRY OVER THE RIVER TEES, AT MIDDLESBROUGH.

DESIGNED BY MR. CHARLES SMITH, ENGINEER, HARTLEPOOL.

(For Description, see Page 62.)

FIG 1.

650' 0"

150' 0"

FIG 2.

FIG. 3.

FIG. 4.

FIG. 5.

Heavy Section.

FIG. 7.

Lattice Bars

FIG. 10.

Heavy Section

3' 9"

3' 9"

Lattice Bars

FIG. 8.

Light Section

Lattice Bars

Lattice Bars

FIG. 9.

Light Section.

JOINT OF CENTRE GIRDER WITH CANTILEVER

FIG. 6.

Lattice Bars on all sides

FIG. 11.

MAIN CROSS BRACE

TIE BARS

TIE ROD

TIE BARS

FIG. 12.

FIG. 13.

GONDOLAS IN THE AIR

In 1873, Charles Smith, a thirty-year-old engineer born in Arbroath in Scotland, was three years into his tenure as Manager and Chief Engineer at Castle Eden Foundry, the Hartlepool marine engineering works of Thomas Richardson & Sons – he would later become a partner in the firm – when he proposed a truly original idea.

Aware that there were numerous river crossings where the construction of conventional high level bridges able to give the required clearance to tall ships would be either impossible or horrendously expensive to build – and despite having no experience whatsoever in bridge design or construction – he proposed what he called his 'Bridge Ferry'.

Two rivers where such challenges would arise were already clearly in his mind – the Clyde and the Tees – and Smith initially worked up his 'bridge ferry' design as a solution to the challenges of bridging the Tees.

The concept was simple – two tall pylons topped at high level by a cantilevered gantry spanning the river, suspended beneath which by long cables would be a platform almost at water level and which could be moved

Opposite: Charles Smith's proposed design for the Tees Bridge, as published in *Engineering*, 25 July 1873.

Below: There is much similarity between the cantilevered design of the Tees Transporter Bridge – seen here in a postcard published shortly after it was completed in 1911 – and Charles Smith's 1873 proposed 'Bridge Ferry' design. The 1911 bridge even crosses the Tees approximately where Smith had proposed building his bridge.

Transporter Bridge, Middlesbrough

JV 70621

Middlesbrough would have to wait 38 years after Charles Smith published his bridge proposal before the River Tees was finally crossed by a transporter bridge in 1911. Amid great pomp and ceremony, the bridge was opened by HRH Prince Arthur of Connaight. This photograph was taken just before the bridge opened to traffic.

from one shore to the other by a system of 'endless ropes' driven by steam-powered winches behind the base of one of the towers.

The suspended platform, carrying both people and railway carriages, would spend much of its travelling time well out of the shipping lanes, ensuring minimal disruption to the considerable amount of traffic which used the river.

As Smith was an accomplished engineer and already responsible for some innovative improvements to the efficiency of steam engines and boilers – particularly those used in steamships – his proposals were pretty much the 'finished article', and his 'bridge ferry' could have been constructed and brought into operation relatively quickly at, he believed, a total construction cost of just £31,192.

He had built in 'back-up' systems – including two steam engines, one in reserve – to ensure the operation of the bridge would not be interrupted by breakdowns. He even calculated the annual operating cost of the bridge as being £960 and suggested that a further £300 per year should be put aside for maintenance. He eventually submitted his initial proposal and technical drawings to the Middlesbrough Ferry Committee – which had oversight of the many small paddle-steamer ferries plying between the shores of the River Tees.

The ferry committee had, just ten years earlier, introduced the paddle-steamer *Progress*, and in 1873 they also introduced the PS *Perseverence*, which was the first vessel large enough to carry horse-drawn vehicles.

They shared the river crossing with numerous licensed rowing boats, but the demand for transport between Middlesbrough – originally known as Port Darlington – on one shore and Port Clarence on the opposite shore just kept growing.

There were already more than a million passenger crossings per year being made when Smith's proposals were submitted to the Ferry Committee and his very detailed description of the proposed bridge was also published as a two-page article in the 25 July 1873 edition of the journal *Engineering*.

'This 'bridge ferry' has been designed by Mr. Charles Smith, Manager of the Hartlepool Iron Works, Hartlepool, and it includes many excellent as well as novel features, which will render it entitled to special attention, not only by the authorities of Middlesbrough, but by those of other places where similar conditions apply.

Of course, in the case of a busy river like the Tees, it is of great importance that any means adopted for accommodating the cross traffic should interfere as little as possible with the carrying on of the traffic up and down stream, and if no other objection existed, there can be no doubt that a tunnel or a high level bridge would afford the best means of attaining this end. But both the tunnel and high level bridge, independent of their high cost, involve the difficulties of the approaches, and in the case of the high level bridge these would, in a flat district like that of Middlesbrough, form a practically insurmountable

Ferdinand Arnodin's cable-stayed and cantilevered Pont Transbordeur at Marseille, completed in 1905, had a span of 165 metres. It was featured on many postcards published thereafter.

156 = MARSEILLE.
Le Bassin de Carénage et le Pont Transbordeur

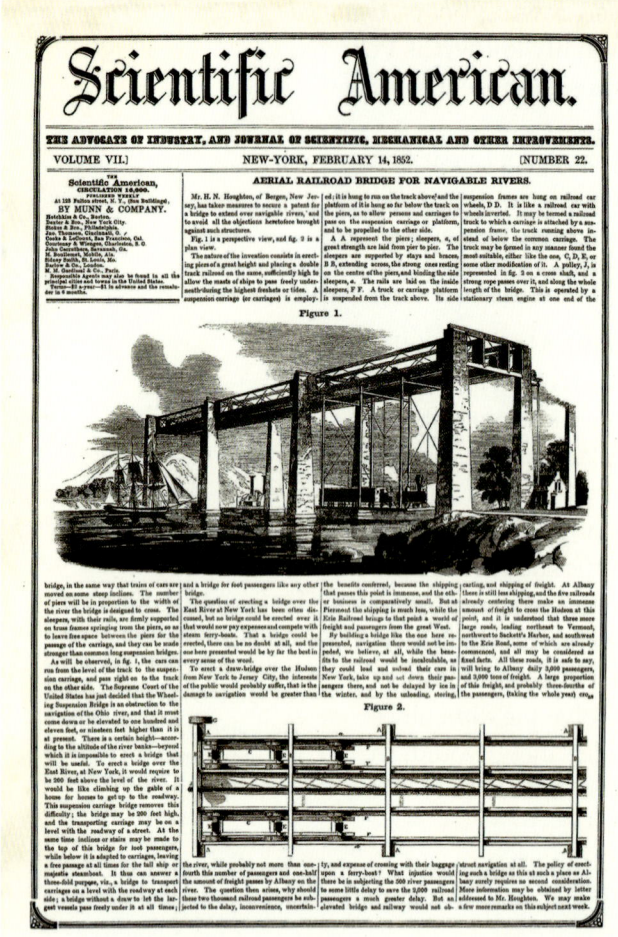

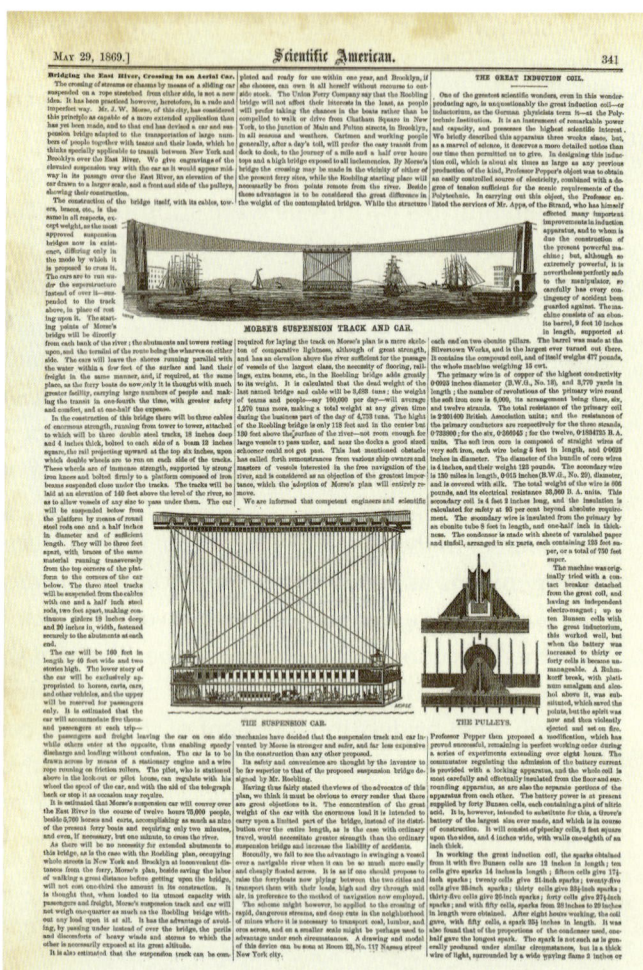

Above left: H.N. Houghton's 1852 proposal for an aerial railway bridge, illustrated in the 14 February 1852 issue of *Scientific American*.

Above right: The Scientific American article which explored J. W. Morse's design for the East River Aerial Bridge, was published in May 1869. A suspension design like this could never have supported such a large and heavy gondola over as wide a span as this, and experts today suggest that a cable-stayed design might have been more likely to give the necessary rigidity.

obstacle, so long as the bridge was used by the traffic passing over it in the ordinary way. Mr. Smith, however, has designed a means of utilising a high level bridge, while maintaining the passengers and goods to be transported at the ordinary level of the river banks, and the means by which he has attained this end we shall be able to explain by reference to our engraving on page 60.'

The article ended by outlining the commercial viability of the project and further underlined just how detailed Smith's proposal had been.

'Mr. Smith estimates the transporting capacity of his 'bridge ferry', even if worked at a speed of but five miles per hour, as amounting to 1,380,000 tons per annum, the working time being taken as but twelve hours per day, while, if worked eighteen hours per day, be take as about 2,000,000 tons. With even half this amount of traffic, or, say, 1,000,000 tons per annum, a charge of but a

little over three farthings per ton would, as he states, provide for a dividend of 6½ per cent. on the estimated capital required, this being independent of the revenue derived from passengers.'

The writer concluded that the idea was so well developed that its construction would be relatively straightforward, and that it had received 'very favourable attention in the Middlesbrough area'.

The final paragraph painted a rosy picture of the future which the writer felt could lie ahead for such a design and echoed Smith's suggestion that:

'Besides Middlesbrough, there are many places—Glasgow, for instance— where such a 'bridge ferry' as we have described would do good service, while we expect to find that Mr. Smith's plans will receive considerable attention on the other side of the Atlantic.'

John Anderson's 1885 design for a railway transporter across the Hudson River was an elaborate structure based in part on Houghton's 1852 ideas. It would have been inherently unstable.

There was widespread published praise for the innovations in his proposal, including support from experts including Sir Benjamin Baker, the architect of the innovative cantilevered Forth Bridge, but despite all that, perhaps the idea was just a bit too revolutionary or, like several other similar proposals, simply ahead of its time.

Although Glasgow council didn't show any interest in the idea, Middlesbrough at least checked out Smith's proposed costings. He claimed his bridge would cost an estimated £31,162, whereas building a new ferry would cost only £2,975, with an additional £9,981 for the necessary approach roads and landing stages.

While Smith articulated his ideas fully in his presentations, he was neither alone nor the first to believe that this novel idea might be the solution to bridging wide rivers where the surrounding landscape precluded the building of conventional crossings.

Some of the earliest were reported by Henry Grattan Tyrrell in his 1912 pamphlet *Transporter Bridges*, including one which, Tyrrell claimed, could transport a full length railway train. His described it thus:

> 'A patent for an aerial railway bridge to cross the East River at New York was granted about 1852 to H.N. Houghton, of Bergen, N.J., who proposed placing a number of heavy stone piers in the river, with truss spans thereon, and a clearance under the spans of 150 to 200 feet for ships. Instead of expensive approaches to a high level bridge, he proposed suspending a moving platform for a double line of railway, and making this platform long enough to carry whole trains of cars. The obstruction which this plan offered to shipping was from the river piers only, the space between them being always open except for the occasional passing of the moving platform.'

Tyrrell had, however, misread Houghton's proposal – the 'double line of railway' in fact referred to the twin tracks from which the proposed gondola would be suspended, rather than twin railway lines *on* the gondola as he reported.

Back in 1852, *Scientific American* had devoted the entire front page of its issue of 14 February to Houghton's proposal in which it is clear that the suggested East River crossing was just one of the suggested applications of his idea.

The illustration was of a stone-built bridge on a much lesser scale than would have been needed to bridge the vast span of the East River, and rather than transporting a full length train, this version carried a locomotive and a single coach and would have thus been of very limited practicality.

According to the report, however, Houghton believed his idea would be the answer to bridging any navigable channel, whatever the span involved.

> 'Mr. H.N. Houghton, of Bergen, New Jersey, has taken measures to secure a patent for a bridge to extend over navigable rivers, and to avoid all the objections heretofore brought against such structures.

Fig. 1 is a perspective view, and fig. 2 is a plan view.

The nature of the invention consists of erecting piers of a great height and placing a double railroad track on the same, sufficiently high to allow the masts of ships to pass freely underneath during the highest freshets or tides.

A suspension carriage (or carriages) is employed; it is hung on the track above and the platform is hung so far below the track as to allow persons and carriages to pass on the suspension carriage or platform, and to be propelled to the other side… A truck or carriage platform is suspended from the track above. Its side suspension frames are hung on railroad car wheels… It is like a railroad car with wheels inverted.'

Houghton proposed that his gondola would run between each pair of towers as it made its way across a river, and as he suggested that traction for the moving platform would come from a stationary steam engine and rope or cable winch at one end of the bridge, there would have been limits to the length of any such bridge. The cost of spanning something as wide as the East River would never have been an economic proposition.

In 1869 J. W. Morse suggested bridging the East River linking New York and Brooklyn with a single suspension span 430 metres in length and with a clearance of 45 metres above water level but his radical – and impractical – idea was never put to the test.

A detailed account of his proposal – 'Bridging the East River, Crossing in an Aerial Car' – was carried in *Scientific American* on 29 May 1869, which reported that the idea of a transporter bridge, while not entirely new:

'…has been practiced however, heretofore, in a rude and imperfect way. Mr. J.W. Morse, of this city, has considered this principle as capable of a more extended application than has yet been made, and to that end has

The journal *The Engineer* published illustrations of four different types of transporter bridges in its issue for 27 March 1908. The profile of the left hand tower of the Marseille bridge, however, is incorrect. Neither of the towers was erected on top of a building.

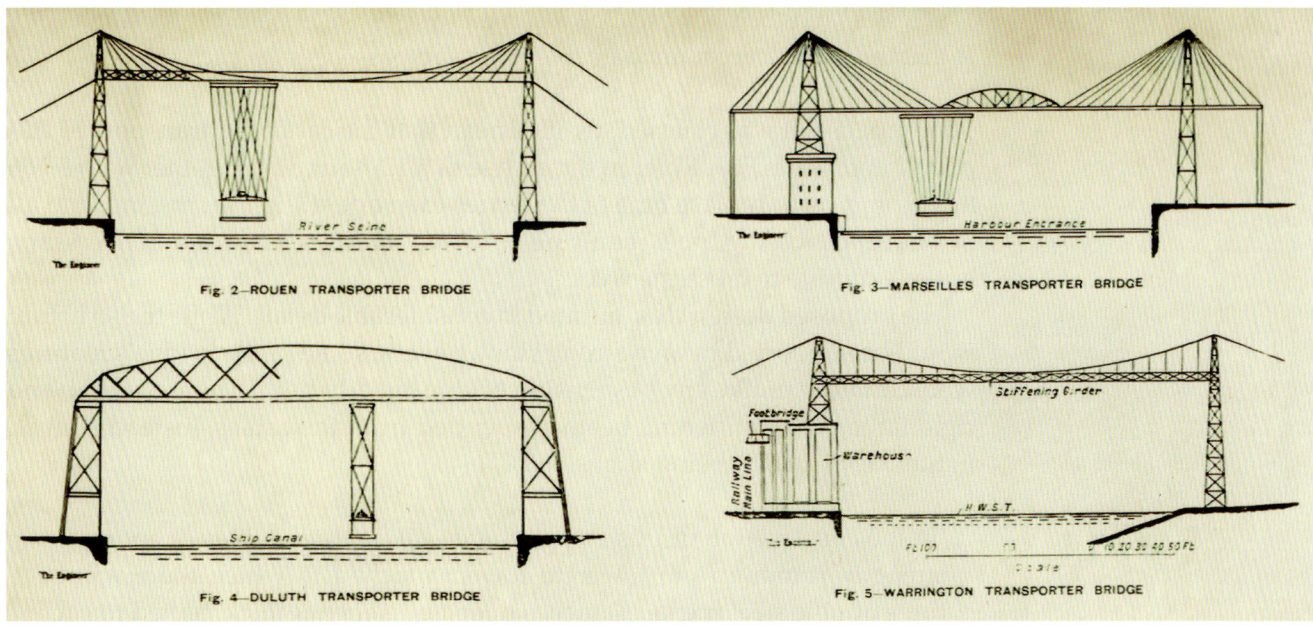

Fig. 2—ROUEN TRANSPORTER BRIDGE

Fig. 3—MARSEILLES TRANSPORTER BRIDGE

Fig. 4—DULUTH TRANSPORTER BRIDGE

Fig. 5—WARRINGTON TRANSPORTER BRIDGE

devised a car and suspension bridge adapted to the transportation of large numbers of people together with teams and their loads, which he thinks specially applicable to transit between New York and Brooklyn over the East River.'

The article was accompanied by illustrations of the projected bridge and its large gondola, and a detailed view of the running gear.

Morse's suggestion was timely, as the writer went on to explain the first plans had been submitted for what would later become known as the Brooklyn Bridge.

'The construction of the bridge itself, with its cables, towers, braces, etc., is the same in all respects, except weight, as the most approved suspension bridges now in existence, differing only in the mode by which it is proposed to cross it. The cars are to run under the superstructure instead of over it— suspended to the track above, in place of resting upon it. The starting point of Morse's bridge will be directly from each bank of the river; the abutments and towers resting upon, and the termini of the route being, the wharves on either side. The car will leave the shores running parallel with the water within a few feet of the surface, and land their freight in the same manner, and, if required, at the same place as the ferry boats do now, only it is thought with much greater facility, carrying large numbers of people and making the transit in one fourth of the time, with greater safety and comfort, and at one half the expense.'

One report suggested that the gondola would cross less than one metre above high water. The gondola itself was planned to measure:

'…160 feet in length by 40 feet wide and two stories high. The lower story of the car will be exclusively appropriated to horses, carts, cars and other vehicles, and the upper will be reserved for passengers only. It is estimated that the car will accommodate five thousand passengers at each trip…'

That capacity, it was claimed, would mean that the car could transport 75,000 people across the East River in the course of a 12-hour day, together with 3,760 horses and carts, replace nine of the ferries then operating the crossing, and all in a normal crossing time 'requiring only two minutes, and even, if necessary, but one minute to cross the river.'

The proposed design was outlined in considerable detail. While the structure would be supported by two strong stone towers, some of the bridge's features were certainly unlike any suspension bridge already built. There would be no rigid lattice-truss stiffening beam suspended from the cables, instead a much lighter construction being suggested.

'In the construction of this bridge there will be three cables of enormous strength, running from tower to tower, attached to which there will be three double steel tracks, 18 inches deep and 4 inches thick, bolted to each side of a beam 12 inches square, the rail projecting upward at the top 6 inches, upon which double wheels are to run on each side of the tracks. These wheels are of immense strength, supported by strong iron knees and bolted firmly to a platform composed of iron beams suspended close under the tracks.'

Morse's proposal was supported by what appeared to be very detailed calculations and figures – including estimates that the average weight the suspension cables would have to support during the working day would be 4,753 tons – and the plan reportedly had the tacit endorsement of a number of un-named engineers. He claimed – unrealistically – that his bridge could be constructed and open to traffic within one year, and would be 'stronger, safer, and far less expensive' than John Augustus Roebling's competing plans.

Not only would it have been impossible to build within Morse's timescale of a year, it would simply have been impossible to build at all.

By the time he published his ideas, plans for the proposed Brooklyn Bridge, designed by Roebling, had already been published, and had certainly not met with universal approval.

The roadway, it was said, at just 40 metres above the river in the middle of the channel and 36 metres near either shore, would be too low to allow the safe passage of clippers into their berths.

Completed in 1915, the Ponte Alexandrino had several structural similarities to the much larger Widnes–Runcorn Bridge completed a decade earlier.

Ponte Alexandrino de Alencar, Bahia do Rio de Janeiro, Brazil

Right: The Ponte Alexandrino had a span of 171 metres between the towers.

Below: Photographed from the air in 1918, the Ponte Alexandrino can be seen on the right of the picture, linking the naval base on the Ilha das Cobras with Rio de Janiero on the mainland.

RIO DE JANEIRO — PONTE ALEXANDRINO, DO ARSENAL Á ILHA DAS COBRAS

Despite devoting two-thirds of a page to Morse's proposal, and a great amount of detail – all supplied by the bridge's proposer one imagines – most of it in apparent support of the idea, *Scientific American* was certainly not entirely convinced that the idea of a transporter bridge could ever be made to work, concluding:

> '...we think it must be obvious to every reader that there are great objections to it. The concentration of the great weight of the car with the enormous load it is intended to carry upon a limited part of the bridge, instead of its distribution over the entire length, as is the case with ordinary travel, would necessitate greater strength than the ordinary suspension bridge and increase the liability to accidents.
>
> Secondly, we fail to see the advantage in swinging a vessel over a navigable river when it can be so much more easily and cheaply floated across. It is as if one should propose to raise the ferryboats now plying between the two cities and transport them with their loads, high and dry, through mid air, in preference to the method of navigation now employed.'

Their scepticism was well founded. Had the bridge ever been built, it would have proved quite useless. By the time the huge gondola car was just one quarter of the way across the East River, the suspension span would have deflected and deformed to such an extend that the gondola would have been up to three metres below the surface of the water.

In the absence of any detailed account as to how the towers and the suspension cables were going to be anchored to the ground, a question mark must hang over whether or not the structure itself could have withstood such deformation.

With the benefit of hindsight, had Morse opted for a cable-stayed design rather than suspension cables, his idea might have had a greater chance of working, considerably reducing the likely stresses on the rails from which the gondola was suspended.

The question must be asked, however, as to why such a respected journal as *Scientific American* would devote so many column inches to an account of a fundamentally flawed proposal. The same accusation, of course, might be made for discussing it on these pages.

So who was J.W. Morse? Extensive research has failed to find any engineer with that name at around the time the article was published in *Scientific American*, but an interesting proposition can be developed around the actual technical details of the proposed bridge, and the illustrations of it. The lack of a rigid main truss or beam, and the effect on the suspended rails of such a large and heavy gondola throw into question whether or not the proposed structure would have been dimensionally stable, perhaps strongly suggesting that the mathematics which underpinned it was not the work of an engineer.

Close examination of the illustrations in *Scientific American* perhaps offers a possible answer. They carry the engraver's name – Morse – and a Joseph

W. Morse was a well-respected engraver based in New York working from the 1840s into the 1860s. Might it be possible that the 'designer' and engraver were one and the same man?

The coincidence of there being an engraver by the name of J.W. Morse making the printing blocks for his namesake's bridge proposal is unlikely. Whoever he was, J.W. Morse clearly foresaw a viable future for the as-yet-untried idea of the transporter bridge.

Had Charles Smith been aware of Morse's proposal made four years before he submitted his ideas to the authorities in Middlesbrough? His cantilevered design was certainly of a much more robust and rigid design than Morse's rails suspended from three steel cables, but was the eventual rejection of his proposal by the Middlesbrough authorities based on the same misgivings that the writer in *Scientific American* had expressed over Morse's?

Despite proposing a more viable design, Smith became the almost-forgotten man in the history of the transporter bridge – a man whose idea was, it seems, well ahead of its time. Whereas J.W. Morse's proposal would never have worked, Smith's potentially viable contribution to the evolution of the transporter bridge was simply put aside.

What Smith's future career might have been will never be known as he was drowned in a swimming accident in Lake Lucerne in 1882 at the age of only 38. While he never saw one of his bridges built, at least he never lived to see others claim recognition for his invention.

With hindsight, Houghton, Morse, Smith and others were visionaries who recognised the world's growing need for an increase in the number and efficiency of river crossings to cope with ever-growing traffic. The number of passengers annually crossing the Tees by ferry, for example, would more than double from its 1873 figure before the Tees Transporter bridge – looking very like Smith's 1873 proposal – was actually built. It was eventually opened in 1911.

According to Tyrrell in *Transporter Bridges*, others had also proposed some ideas along broadly similar lines, including a proposal for a transporter bridge for London.

Five years later, an elaborate plan for a transporter bridge over the Thames was prepared by L. Mills and A. Twyman of North Shields, with a centre opening 200 feet in width and 80 feet high. The upper platform, reached by elevators in the towers, was to have provision for pedestrian travel, so that foot passengers could cross at all times.

What Tyrrell described as 'an elaborate plan' was, in fact, no more than a letter by Mills and Twyman – who gave their address as New Quay, North Shields, although no record of them has yet been found in contemporary trade directories.

The letter was written on 14 March 1878 to the journal *Engineering* and appeared in the issue for 29 March under the heading 'Proposed Movable Bridge over the Thames, Designed by Messrs L. Mills and A. Twyman, North Shields' and contained an outline proposal for their bridge. It was reproduced

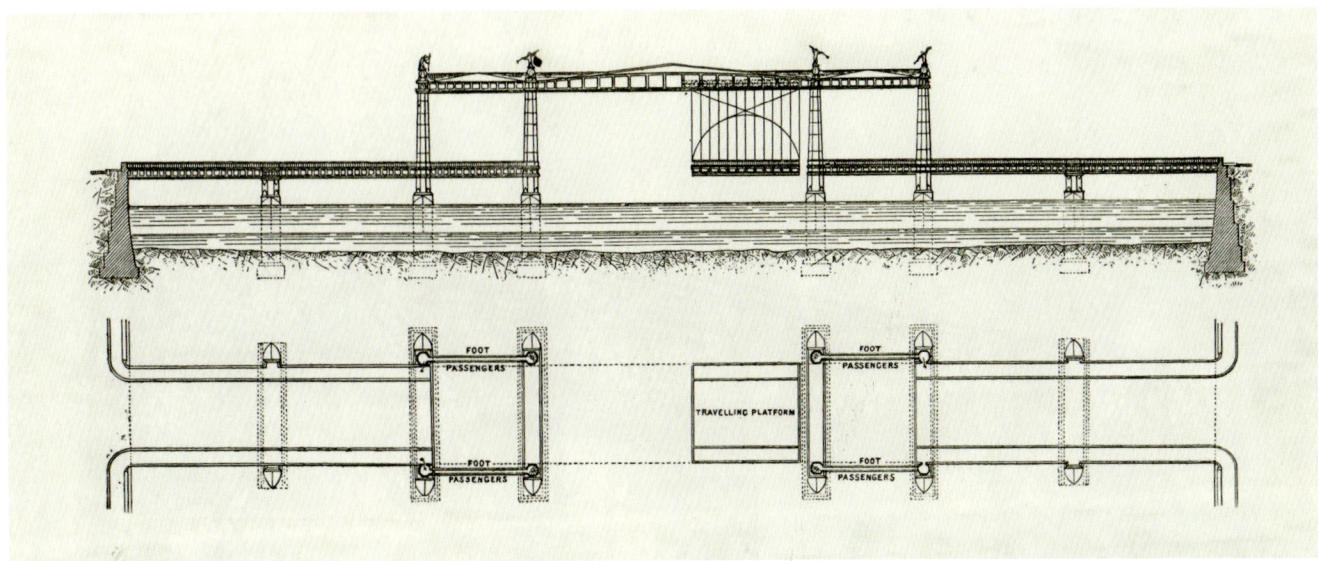

This illustration of Mills and Twyman's proposed Thames transporter bridge was published in the 29 March 1878 issue of the journal *Engineering*. It is clear from the drawing that the 'travelling platform' was designed to dock fully within the towers at either end of its traverse, thus leaving a wide clear shipping lane of the height and width dimensions quoted by Tyrrell. The central 'transporter' section of the bridge has a lot in common with H.N. Houghton's 1852 proposal published in *Scientific American*.

Opening Day on the Pont Transbordeur in Nantes, 1903.

Work nearing completion on the Osten transporter bridge in Germany in 1909. The bridge spans the River Oste in Cuxhaven, Lower Saxony. The asymetrical design of the bridge is unique.

a few pages after the illustration of the bridge that they had submitted, and on the face of it appears to have been anything but elaborate – a simple but robust structure with some innovative features – and one which might well have worked.

'At the level of the roadway and from the opposite shores, piers are constructed by any ordinary method that may be applicable to such a distance as will leave a free and uninterrupted passage for shipping of any reasonable width, and across this, the land traffic is transported by means of a travelling platform suspended from girders at such superior elevation as will permit the passage of masted vessels; this platform when at either end of its traverse, is housed between the towers which support the upper girders, and an uninterrupted road for foot-passengers may be secured, if necessary, by spiral stairs in the towers. The main roadway, if advantageous, may rise towards the centre of the river by an easy gradient for the better passage of smaller craft between the piers.

The travelling platform is suspended by links or other suitable means from a frame carried upon wheels, the whole forming a carriage which may be moved by hydraulic power, the requisite machinery being placed in the towers at the ends of the superstructure.'

Mills and Twyman estimated that if their bridge operated for 20 hours per day, crossing at 10-minute intervals, it could carry around 400 passengers per crossing, and nearly 3,000 vehicles a day. It was, of course, never built and appears not to have been mentioned again in print. Despite the drawings and predicted capacity, it was very much a 'work in progress', with the two men failing to even identify which part of the Thames they proposed bridging. Their idea joined the growing list of highly innovative ways of getting people and vehicles across busy shipping lanes which never got off the drawing board.

Early twentieth-century writers describing the construction of transporter bridges across Europe would all make reference to J.W. Morse's and Charles Smith's original proposals, but not to Houghton, Mills, Twyman and others, and, as *The Engineer* correctly predicted in March 1908:

'...but when the history of the recent remarkable exploitation of the transporter bridge principle comes to be written, due credit will be accord to M.F. Arnodin of Chateau Neuf sur Loire, who, realising the possibilities of the subject, made it his particular study, erecting the first transporter bridge at Portugalete, a small port just below Bilbao in Spain.'

While Morse's proposal can be easily dismissed, and Smith never saw his more viable proposition realised, it is to Alberto Palacio and Ferdinand Arnodin that the credit must go for designing and overseeing the construction between 1890 and 1893 of the first transporter bridge in the world. Several

Little is known about the history of the Kiel bridge, but it featured in a number of postcards in its relatively short existence. This card was posted in July 1913.

Kiel. Die große Schwebefähre der Kaiserlichen Werft.

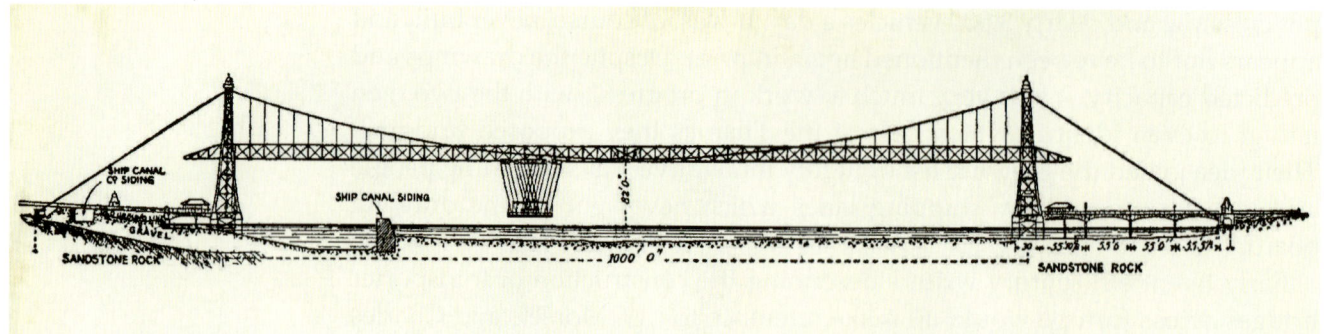

Britain's first transporter bridge crossed the Mersey from Runcorn across the Manchester Ship Canal then across the river itself to Widnes on the north shore.

other designers and engineers on both sides of the Atlantic – some of whom remain unrecognised – would follow them over the years.

Arnodin designed and built the most – a total of nine – and jointly patented the idea with Palacio in 1887, with whom he worked on the Puente Viscaya in Portugalete, Bilbao, which has been listed as a World Heritage Site by UNESCO since 2006.

During the years when such bridges were being built, they aroused considerable local and national interest – their innovative design even meriting an entry in the 1911 edition of the *Encyclopaedia Britannica*:

'This new type of bridge consists of a high level bridge from which is suspended a car at a low level. The car receives the traffic and conveys it across the river, being caused to travel by electric machinery on the high level bridge. Bridges of this type have been erected at Portugalete, Bizerta, Rouen, Rochefort and more recently across the Mersey between the towns of Widnes and Runcorn.

The Runcorn bridge crosses the Manchester Ship Canal and the Mersey in one span of 1000 ft., and four approach spans of 55½ ft. on one side and one span on the other. The low-level approach roadways are 35 ft. wide with footpaths 6 ft. wide on each side. The supporting structure is a cable suspension bridge with stiffening girders. A car is suspended from the bridge, carried by a trolley running on the underside of the stiffening girders, the car being propelled electrically from one side to the other. The underside of the stiffening girder is 82 ft. above the river. The car is 55 ft. long by 24½ ft. wide. The electric motors are under the control of the driver in a cabin on the car. The trolley is an articulated frame 77 ft. long in five sections coupled together with pins. To this are fixed the bearings of the running wheels, fourteen on each side. There are two steel-clad series-wound motors of 36 B.H.P. For a test load of 120 tons the tractive force is 70 lb per ton, which is sufficient for acceleration, and maintaining speed against wind pressure. The brakes are magnetic, with auxiliary handbrakes. Electricity is obtained by two gas engines (one spare) each of 75 B.H.P.'

Ironically, they were described in the encyclopaedia as being a 'new type of bridge' just as their popularity was waning. Relatively few were built – with the rise of motor transport, their limited capacity proved their undoing and those which survive have become iconic structures.

Above left: An early postcard of the world's second transporter bridge, opened in 1898 across the harbour mouth at Bizerte, Tunisia. It lasted just eleven years before being dismantled when the naval base was being expanded. It was subsequently rebuilt at Brest in Brittany, France.

Above right: The rigid suspension frame and the integral motors and cable drive used on America's only transporter bridge were unique. The bridge – which entered service in 1905 – crossed the harbour mouth at Duluth Minnesota.

But why are they called 'transporter bridges'? There are two schools of thought: the first, that they transported people and vehicles across open spaces, is less persuasive than the other. Ferdinand Arnodin, being French, used the term 'transbordeur' to describe them – the word meaning 'across the divide', or 'from one side to the other', and that 'transporter' is simply a corruption of that. The Spanish form 'transdordador' can also be translated as 'shuttle'. Either of these could be the origin.

Engineers differ slightly in what their definition of a transporter bridge actually is, but by the most common definition – a single span bridge carrying

Hochbrücke
bei Rendsburg mit Schwebefähre

Above: The gondola starting out on its crossing of the Kiel Canal just after a warship has passed beneath the Rendsburg Bridge. The bridge was completed in 1913 and this card dates from around that time. The transporter is out of service for repairs at the time of writing, but this remains the only combined railway/transporter bridge ever built.

S. M. S. „Hohenzollern" die Hochbrücke b. Rendsburg passierend

Right: The German Kaiser's 1893 steam yacht *Hohenzollern* passing beneath the Rendsburg Bridge at its opening in 1913.

a low level gondola across a river – there were probably no more than 21 completed in the entire world.

That definition rules out several which might more obviously be described as aerial cableways, including a Russian bridge said to have been built in Stalingrad, now Volgograd, in 1955 with a claimed span of 874 metres!

There is some doubt that it ever existed, as there appear to be no records of its construction, no photographs of it, and no record of its demolition. A single span of such a length would be well beyond the limits of engineering even today, and no contemporary maps of Stalingrad appear to show any sign of such a structure.

There were, understandably, several different design solutions to the construction of such bridges. At its simplest, a transporter bridge consisted of two tall pylons carrying a lattice steel boom or 'stiffening beam' across the river which carried the transporter mechanism. Considering how few were built, those 21 bridges were designed using five quite different sets of engineering principles.

Some were simple lattice/truss structures, others were cantilevered while several were effectively suspension bridges. Two were cantilevered and cable-stayed and five others – the Viscaya and Rouen bridges, the Bizerta Bridge later rebuilt in Brest, the Rochefort–Matrou bridge and the Newport Transporter Bridge in South Wales – were built using Palacio and Arnodin's original hybrid suspension/cable-stayed design.

According to Tyrell's pamphlet *Transporter Bridges*, there was a 22nd bridge, in Tangier, but research has so far failed to unearth any solid information about its location, size, appearance or designer – although Tyrrell was confident enough of his sources to attribute its design to Ferdinand Arnodin.

Charles Smith's original concept proposed steel cables drawn by steam-powered winches to move the gondola across the river. The majority of those that were built were driven by electricity from the outset – some using steel cables and winches, others using DC motors driving the bogies beneath which the gondola was slung.

Just eight survive – one each in Spain, France and Argentina, two in Germany and three in the UK. Until the 1960s, Britain still had all five of its transporters. That there are still so many of these bridges standing – six still working and a seventh shortly to join them – is down to preservation groups and 'friends' groups who have lobbied hard to get protection for them.

Spain has the oldest – the 164m Viscaya Bridge – still in regular use, albeit largely rebuilt after the Spanish Civil War and now with a replacement beam and a modern gondola hanging from its cables.

The 104m Puente Transbordador Nicolás Avellaneda in Buenos Aires, disused since 1960, has recently been re-opened after a major restoration.

Like the 84m Osten Bridge in Germany, the 121m Duluth Bridge in Minnesota, and the 180m Ponte Alexandrino in Rio de Janiero, the gondola was carried below a steel frame rather than suspended on cables. This enhanced the stability of the gondola on short span bridges as it was moved across the river. To achieve the required stability with cable suspension, the travelling frame ideally had to be around one and a half to two times the length of the gondola itself, thus splaying the cables out as they went up towards the carriage.

The further their spread, the more stable the ride, especially when there was a wind. *The Engineer* in March 1908 drew attention to those very challenges facing the designers of transporter bridge gondolas.

'The design of the car suspender is another and very important matter; the reason why will be best understood when it is reflected that most transporter bridges do an important traffic in conveying horse-drawn vehicles with the horses in the shafts. To avoid exciting the horses, the car must start up very gradually without shock or oscillation; when in mid-stream the suspender must be able to resist any tendency to swing sidewards caused by the wind; and, finally, a tendency which is perhaps the worst of all, that of the car to swing, not only longitudinally, but diagonally, when slowing up at its landing stage, especially after it has been necessary to apply the brakes, must be most carefully guarded against. Car oscillation may be avoided in two ways, first by careful design of the suspender, and, secondly, by a properly thought out trolley drive. In some districts the form of suspender is, to a great extent, determined by the shape of the car, which in turn is decided by the shape of the vehicles it has to accommodate; for instance in France most carts are long and narrow, and the horses are harnessed in tandem, thus creating a class of traffic which obviously calls for a long car. A long car being adopted, its satisfactory suspension from the trolley calls for no great thought, as the cables can be spread out over a very long base, and in this way longitudinal oscillations can be kept within unimportant limits. If, however, a short car would accommodate the traffic of the district, then the best form of suspension is by a rigid frame such as that adopted at Duluth. According to Mr. Turner, designer of the Duluth Bridge, its car has only been known to rock ½in. in a 60-mile gale.'

The Duluth 'Aerial Bridge' was the brainchild of the City Engineer, Thomas McGilvray, who commissioned structural engineer Claude A.P. Turner of the American Bridge Company to draw up plans for his unusual and original design in 1899.

Intriguingly, given that the two bridges have little in common visually, McGilvray claimed that Turner's design was based on Ferdinand Arnodin's Rouen bridge which had been opened that same year. 'Inspired by' rather than 'based on' might be more appropriate. There does, however, seem to be evidence in support of the claim that Arnodin offered his services as a consultant during the design and development stage.

While it looked very different to the Rouen 'transbordeur', McGilvray and Turner's drive system was clearly inspired by Arnodin's innovative solution for the French bridge – a break with his previous practice – where he had mounted the winch on top of the gondola. McGilvray installed

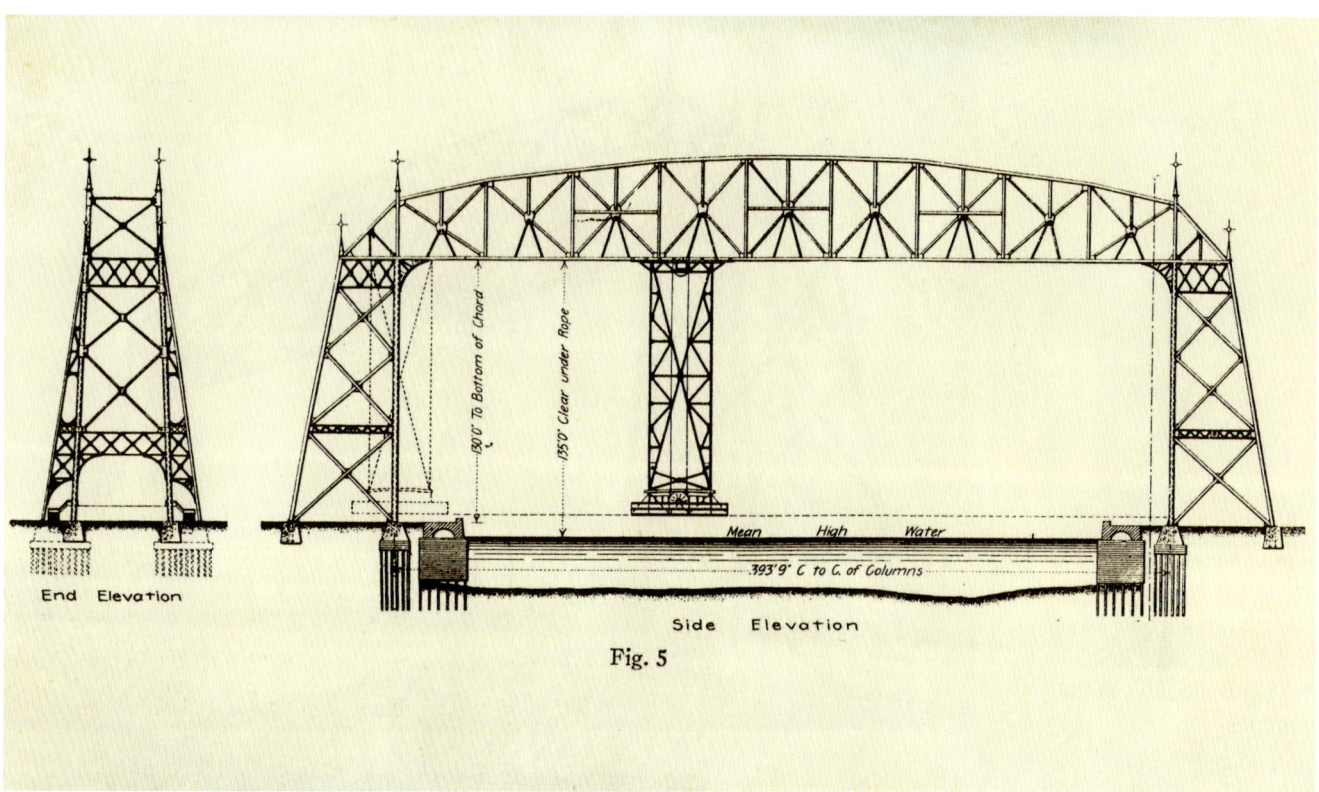

End Elevation

Side Elevation

Fig. 5

two electric winches inside the gondola cabins on his bridge, winding and unwinding steel cables as the deck moved across the harbour mouth.

Work on building the bridge did not start until 1901 and took four years, during which time the plans were modified on a number of occasions, the main contractors were changed, and what opened to traffic in 1905 was a much simpler structure than had been suggested by McGilvray's original drawings. Despite the abandonment of some elements of McGilvray's original concept, it was still revolutionary. Under the heading 'An American Transporter Bridge', *The Engineer* had carried a highly detailed description of the system in its issue for 2 May 1902. While work on actually constructing the bridge had not yet progressed very far when the article was written, detailed plans for it had already been published.

'The transporter type of bridge has been suggested for several points in the United States, but the first place where it has been actually adopted is at Duluth, the great port at the head of Lake Superior…

…Two steel towers on concrete piers will support a pair of riveted steel trusses 394ft. long, 34ft. apart, the lower booms being 135ft. above high water. Trolley trucks running on the lower booms carry riveted and braced hangers, the lower ends of which carry the car. The car is proportioned to carry a tramcar, weighing 21 tons, and to have the remainder of its platform loaded with 100lb. per square foot. The trusses are designed for a dead load of 330 tons and a live load of 120 tons,

Elevation of the Duluth Aerial Ferry Bridge, published in Henry Grattan Tyrrell's pamphlet *Transporter Bridges* in 1912.

The Duluth Bridge's winches and motors were, uniquely, built into the sides of the gondola.

AERIAL BRIDGE, ENTRANCE TO DULUTH-SUPERIOR HARBOR.

AERIAL BRIDGE OVER SHIP CANAL, DULUTH, MINN.
Only one of its kind in the world; clear span, 400 ft.; clear height 136 ft.; total height above water 186 ft.; size of car 34x50 ft.; capacity of car, 125,000 lbs.; motive power, electricity; speed, 4 miles per hour. | PULL ☞ and Return Slowly

with an assumed wind load of 50lb. per square foot on all exposed areas including the car and hanger.'

The gondola was to measure 10m by 16m, with rails and overhead wires for tramcars, but those features are not evident in any surviving photographs of the bridge, suggesting that they were amongst the many changes implemented during construction. Indeed, the approach ramps appear to have a gradient which would have been difficult for trams to negotiate. But it was McGilvray's variation on Arnodin's winch-drive system which marked the Duluth bridge as unusual.

'The bridge car will be operated by electric motors, two of 50 horse-power each being provided, one for use in reserve. The ordinary speed will be four miles an hour, but higher speeds are possible. The operations will be governed by a controller on the car. The power will be applied by means of a 1in. wire rope in the lower boom of the trusses, the rope being supported by automatic hangers, which will swing aside to let the trolleys pass and then drop back to catch the trailing end of the rope. This rope runs through the lower boom of the trusses, passes down through the trolleys and hangers to the car, around a 9ft. friction drum, 24in. wide, on the car – driven by the 50 horse-power motor – and then up again to the boom. The ends of the rope are anchored in the towers. Provision is made for operating the friction drum by hand in case of a failure of the power.'

The slider on this novelty souvenir postcard – published in 1908 and designed by A.J. Hersey for the Duluth Card Company – moved the gondola across the harbour mouth. The caption claimed that the bridge was the 'Only one of its kind in the world.' On a later version that was changed to 'Second of its kind in the world.' In fact it was the world's sixth transporter bridge.

One of a number of silver teaspoons issued to mark the Duluth Bridge's opening in 1905.

The Duluth bridge as it looks today. It was converted to a lift bridge in 1930 by raising the towers and inserting the additional superstructure and lift mechanism.

To commemorate the opening, a range of souvenirs was produced, including Sterling Silver teaspoons with the moulded image of the bridge on the bowl.

Duluth's time as a transporter bridge lasted little more than 20 years as traffic demand outstripped capacity, being converted to a lift bridge in 1929 – using the original towers but with their height increased by almost 50 per cent.

Winches to raise and lower the roadway were installed in each tower, the roadway at full elevation being the same height as the transporter bridge's main truss. It has now been operating as a lift bridge for almost 90 years and remains in use today.

Plans for Germany's Osten–Hemmoor Bridge were first drawn up in 1902, and by 1903 drawings had been produced showing its anticipated design. Presumably for geological reasons, what was actually constructed was slightly different from those first drawings, as the 1908 plans clearly demonstrate.

With an extension to one end of the stiffening beam, the modified design had the effect of allowing the gondola to progress beyond the tower on the north side of the Oste River.

The tower design itself required considerable modification to effect that change – the original intention had been to use what were, effectively, single towers with the gondola coming to rest on the river side of each at either end of its traverse. Requiring the gondola to pass beyond that point meant the adoption of a four-pier design as with most other transporter bridges.

While this was probably the first 'lightweight' transporter bridge, the design of the towers and stiffening beam was not new and appear based on the loading gantries which were already widely known and used to transport

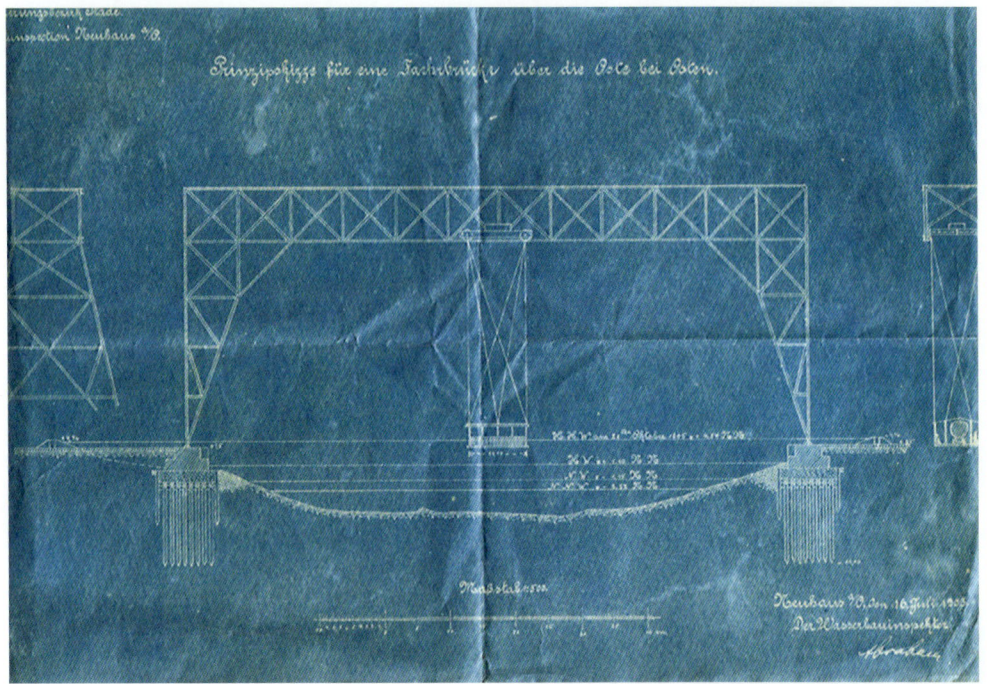

The 1903 blueprint for the Osten–Hemmoor bridge over the Oste River near Cuxhaven in Lower Saxony, Germany.

A postcard view, c.1905, of the Pont Transbordeur at Brive in Corrèze, typical of transporter mechanisms widely used for moving materials around factory yards – in this case moving coal from yard to railway.

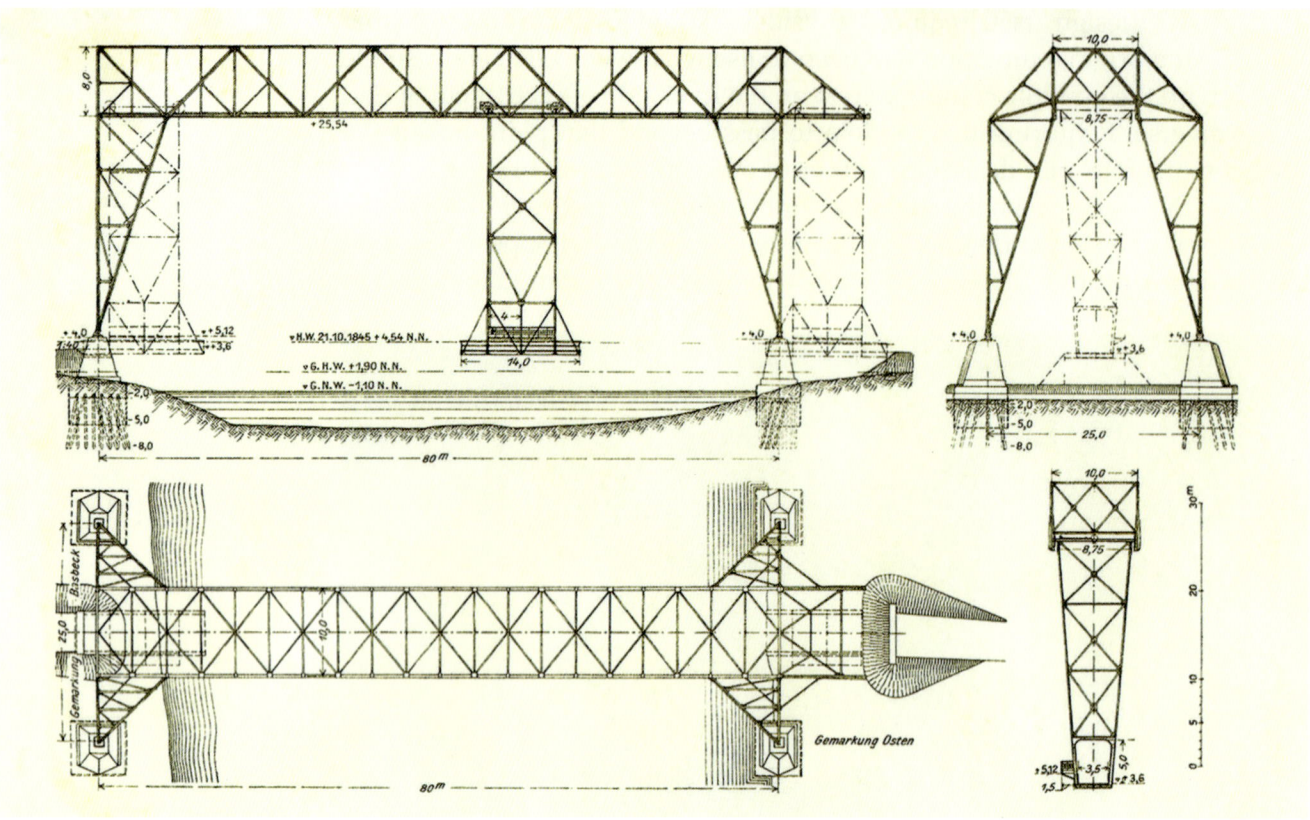

The Osten–Hemmoor bridge was actually constructed between 1908 and 1909 to a slightly modified design. This drawing dates from 1908.

The Osten–Hemmoor bridge as it is today.

Above and overleaf: Two views of the most recent gondola on the Rendsburg Bridge, photographed from a small vessel making its way along the Kiel Canal. This gondola was a much more substantial structure than the original, but nonetheless was severely damaged when struck by a passing ship in 2016.

materials across railway goods yards, industrial manufacturing sites and colliery yards. They were also often referred to as 'ponts transbordeurs' on contemporary postcards.

The relatively short span of the Osten bridge – at just 80 metres – and the strong steel lattice construction of the stiffening beam, enabled the construction of a bridge which required no additional cable anchors. The rigid-frame suspension system used for the gondola made it unique amongst Europe's transporters, all others using cable suspension.

When built, the gondola was powered by DC motors supplied by MAN Werk Gustavsburg and Allgemeine Elektricitäts-Gesellschaft (AEG), but by the 1920s the DC system had been replaced by a much more reliable set of three-phase motors. The bridge opened in 1909 and remained in daily service until 1974, since when it has been operated as a seasonal attraction.

The other survival in Germany is the 140m Rendsburger Hochbrücke – Rendsburg Transporter Bridge – across the Kiel Canal, which uses the more conventional setup of a gondola suspended on cables.

In this case the transporter was suspended below a railway bridge – the only such construction in the world. It was designed by Frederick Voss, took two years to build, and was opened in 1913. It, too, is still normally operational, but the gondola is currently under repair.

The bridge replaced a swing bridge which had previously crossed the canal and, to get the required height – bearing in mind that a 1:150 gradient was considered the maximum viable incline for a train, and that the track had to be raised a total of more than 36 metres to give the required height clearance over the canal – a total of more than 7.5 kilometres of embankments and bridgeworks had to be constructed.

The 140m central span of the bridge, beneath which the gondola is suspended, is just a small fraction of the viaduct's total length of over 2.5 kilometres.

Vehicles and pedestrians leaving the gondola on the Puente Nicholàs Avellaneda in Buenos Aires in 1928. Despite some sources suggesting that the bridge could carry tramcars, there is no evidence of rails.

The crowded gondola of the Puente Nicolás Avellaneda in the late 1920s.

In the 1930s, the Puente Nicolás Avellaneda crossed a busy channel and bustling docks.

Bridge operations were suspended in January 2016 after the gondola was severely damaged during a night-time collision with a passing ship. The gondola was dismounted in March 2016 and, at the time of writing, work is still underway assessing whether or not the bridge can be returned to regular service.

Even if repairs can be effected, however, it seems unlikely that the transporter – which provided an essential vehicle and pedestrian link between Osterrönfeld and Rendsburg – will be operational again anytime soon.

The Puente Nicolás Avellaneda in Buenos Aires was designed and built by the British-owned Buenos Aires Great Southern Railway and opened in 1914. Its construction, however, was a joint Argentine/British undertaking, with the steelwork prefabricated in Brierley – then in Staffordshire but now in the West Midlands – at the Earl of Dudley's Round Oak Iron & Steel Works. It was shipped to Argentina in sections where it was erected by local contractors on to 24m deep concrete foundation piers constructed by the port authorities.

The official opening of the bridge was scheduled for 8 March 1914 – and commemorative medals were struck for the occasion – but problems with the drive delayed the event until 30 May. The surviving gold, silver and bronze medals however – which were issued to guests at the opening ceremony dependent upon their perceived importance – had already been minted and therefore carried the original opening date.

Unused since the 1960s, the bridge has been restored prior to returning it to service as a tourist attraction. The restoration has included a new winch-house, the installation of new motors and reconstruction of the gondola. It re-opened in 2018 after a 160M peso (£8M) restoration.

Buenos Aires once had three transporters, all crossing the narrow Riachuelo-Matanza river and all named after former presidents of Argentina. The Transbordador Sáenz Penã was completed in 1913 to a much simpler design, and the almost identical Transbordador Justo José de Urquiza two years later.

While several photographs have been located of the Puente Transbordador Justo José de Urquiza, the Transbordador Sáenz Penã largely escaped the notice of photographers, appearing in very few photographs. The bridges were demolished in 1965 and 1968 respectively.

One might be forgiven for thinking there were even more – the Puente Nicolás Avellaneda had several different names including 'Puente Transbordador de La Boca', 'Antiguo Puente Nicolás Avellaneda' and the 'Transbordador del Riachuelo'.

When a road bridge was built alongside it in 1940, it too was named after the former president, giving rise to suggestions, still repeated by quite a number of sources today, that the Puente Nicolás Avellaneda was a single combined road bridge and transporter bridge rather than two separate structures located just a short distance apart. Its similarity to the Duluth bridge led some sources to suggest that it was based on the American bridge, but the Puente Nicolás Avellaneda was designed and engineered in Britain.

The gold, silver and bronze medals marking the opening of the Puente Nicolás Avellaneda all bear the date of 8 March 1914, having been struck before the opening was delayed due to technical problems.

Above: The Puente Nicolás Avellaneda, photographed in spring 2017. It was re-opened just a few months later.

Right: The Puente Nicolás Avellaneda part way through its restoration. When this photograph was taken, the new winch-house was nearing completion. Behind the bridge, its replacement – a high level road bridge with lifting central section – was also named the Puente Nicolás Avellaneda, requiring the transporter to change its name and become known as the Antiguo Puente Nicolás Avellaneda.

The 1915 Puente Transbordador Justo José de Urquiza, was dismantled in 1968.

The Puente Transbordador Sàenz Penã from a view of the docks and river published in the1930s.

A proposal for a bridge at Ostend was published in Brussels in 1906 by Presses de Alex. Gielen in an illustrated booklet *Ostende et ses merveilles: le pont-transbordeur.*

If the fanciful illustration on the cover of Gielen's booklet is to be believed, this was to have tall castellated towers on top of which a telegraph station could be built, not only speeding up cross-river traffic, but also improving radio communications both on and off shore.

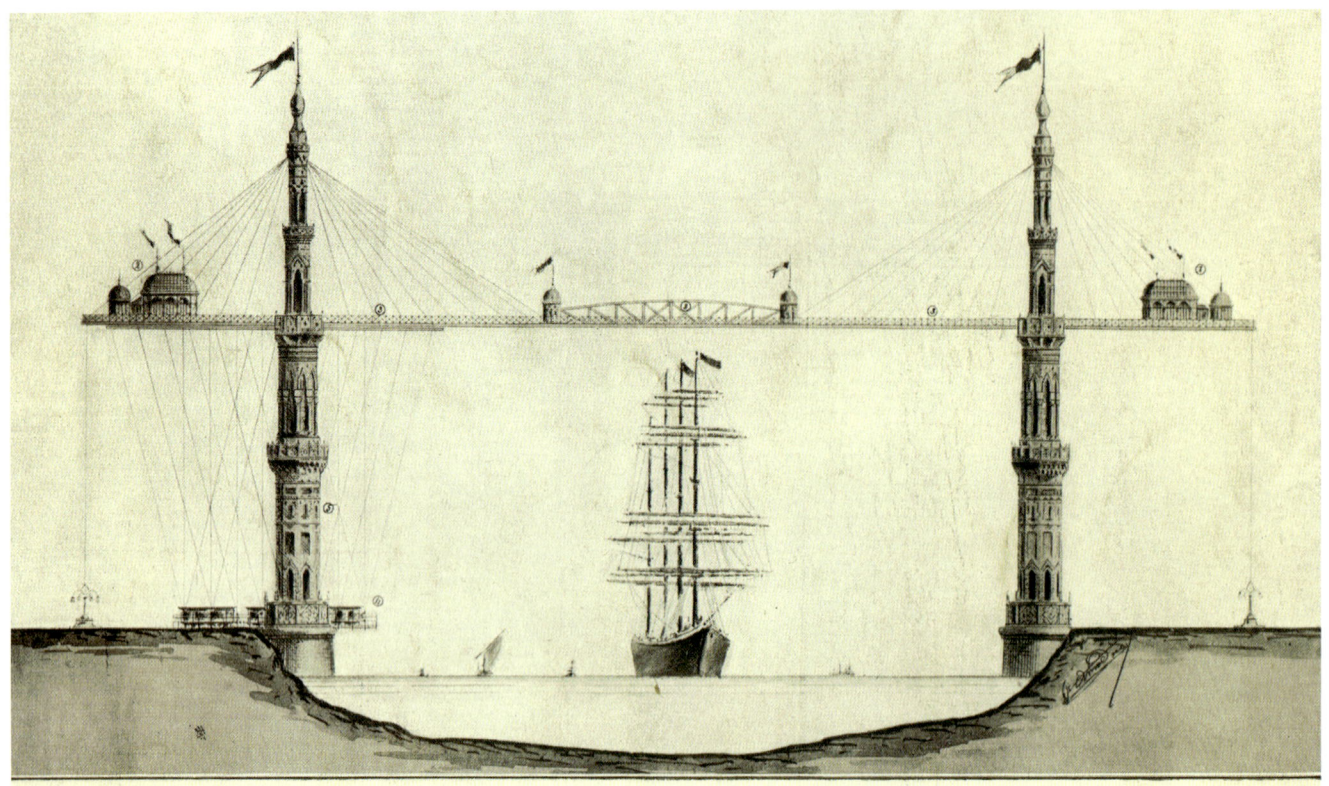

ASPECT GÉNÉRAL D'UN PONT - TRANSBORDEUR

The title page illustration from the booklet *Ostende et ses Merveilles: Le Pont Transbordeur*, published in 1906 by Presses de Alex. Gaelen in Brussels, showing a fanciful elevation of the bridge which might be built.

The main beam was to have been 50m above high water level. with a length of 200m. The gondola, driven by electricity, was to have been 30m by 8m, and in addition to three tram cars and vehicles, it would be able to carry 240 passengers on each journey. Sadly, like all the others, that too never got off the drawing board.

Several local newspapers attempted to justify the construction of the bridge. A translation of a report from the Chamber of Representatives in the Belgian Parliament published in *Le Courrier d'Ostende* on 25 August 1907 reads:

'Unanimously, they expressed a vow in favour of the establishment by the State of a ferry bridge. *Le Courrier d'Ostende* supports this wish, because the need for a rapid and regular means of communication is truly indispensable. The proof is that the population is not increasing at all, that property has diminished in value and that the complaints are increasingly intense.'

A hugely ambitious plan was published by the Southport and Lytham Tramroad Company in 1898 to link the two Lancashire towns using a 'conveyor bridge' across the Ribble Estuary, and the project actually got as far as a draft bill being published in *The London Gazette* on 25 November 1898. The bill envisaged:

'A conveyor bridge across the River Ribble in the townships of North Meols and Lytham, commencing in the township of North Meols at the termination of the last mentioned tramroad, and terminating in the township of Lytham,

at or near the western termination of the northern training wall of the said river, such bridge being intended for the purpose of conveying tramcars, carriages, passengers, animals, and goods by means of a platform suspended from the bridge and worked by suitable machinery.'

Two further bills radically altered the proposed route between Southport and the bridge, changed the nature of the bridge itself, and extended the time-frame within which it must be built. So different were the proposals that they might reasonably be considered as two quite separate projects, although designed by the same man, and with the same intended purpose.

The first one was the more ambitious – and thus the less likely to have ever been practicable. Suggested by John T. Wood, one of the engineers responsible for the Widnes–Runcorn bridge, it proposed a 5 kilometre tramway across marshlands to a docking station where it would meet a Volk's Electric Car similar to – but very much larger than – the one which had been operating between Brighton and Rottingdean since 1896.

The tramcar would then be run on to the Volk Spider Car – which would run on pairs of rails laid on the seabed for the hazardous 3.5 kilometre journey across the estuary where the tidal rise and fall could reach 5 metres – out towards the central navigable channel. It would then arrive at a second docking station where the tramcar would run on to the gondola of the 'Conveyor Bridge'.

Once across the bridge, the tramcar would then run along an 800 metres cast-iron pier before accessing the conventional tramlines to Lytham.

That plan had been abandoned by 1901, as indeed was the proposal for the Volk's Car running out from the south shore.

The Southport and Lytham Tramroad Company's original proposal for crossing the Ribble would have used a very large version of Magnus Volk's electric-powered 'Spider Car' travelling on rails on the seabed for part of the crossing. The original had been successfuly brought into service between Brighton and Rottingdean in Sussex in 1896. Scaling it up to meet the very challenging requirements of the Ribble estuary was not initially seen as an insurmountable challenge. The Brighton experiment, however, had been abandoned long before funding for the Ribble crossing had been achieved.

THE PROPOSED HAYLING ISLAND "AERIAL BRIDGE."

The Evening News in Portsmouth carried an enthusiastic account of the proposed Hayling Bridge in June 1903.

The second plan was for a revised route crossing the river about 3 kilometres further upstream towards Preston and requiring the construction of an 800 metre cast-iron pier running out from the south shore to the bridge, reaching the bridge at a height of 2.5m above high water level. It too never got beyond the drawing board.

In both proposals, the gondola would have been suspended by rigid rods from the travelling frame rather than the cables used on most other transporter bridges in use or being built at the time. This was to counter any oscillation caused by strong winds blowing up the estuary.

As to cost, the transporter bridge was estimated at £53,154, about one sixth of the total proposed cost of the entire tramway.

As successive dates passed, an 'Abandonment Act' was tabled in 1909 as the idea had failed to raise funding.

There were also proposals for a transporter bridge be built linking Hayling Island and Portsea, and despite the fact that this proposal seemed to have a great deal of local support, it too joined the growing list of bridges which were never built. The local Portsmouth newspaper, *The Evening News*, carried a detailed report in its issue of 3 April 1903:

'Today we are enabled to give an illustration of the Traveller Suspension Bridge which it is proposed to construct across the Langstone Harbour channel to Hayling Island. The bridge is a portion of an important scheme for the development of the island which has accorded the hearty approval of

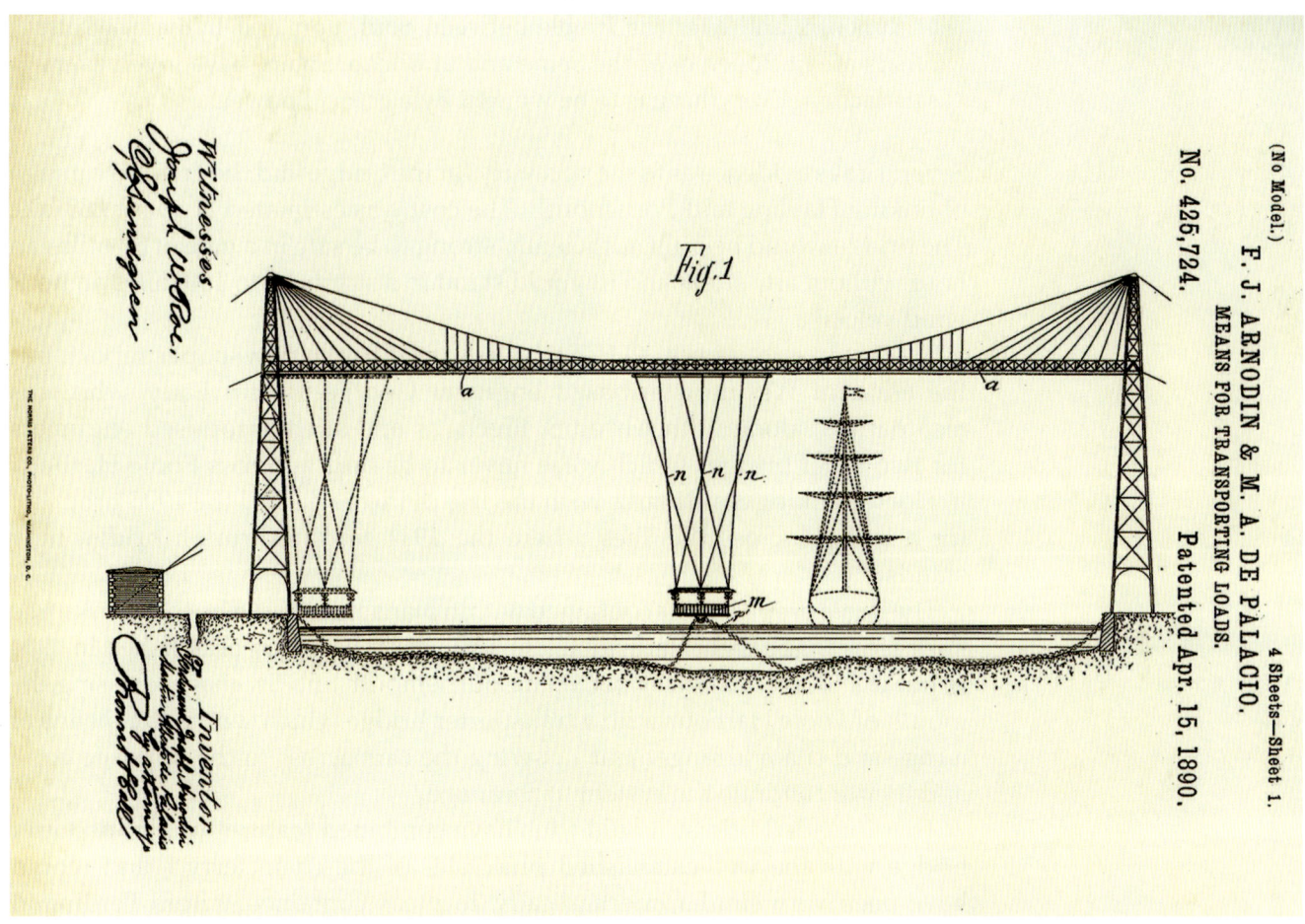

The chain drive system that Eady and Frech planned for the proposed transporter bridge across the mouth of Poole Harbour, was first suggested in Arnodin and Palacio's US Patent of 1900 – this illustration comes from that Patent Specification. The same basic drawing had been used in their 1887 British Patent, but without the chain drive.

a large and influential meeting of landowners and residents held at the Royal Hotel, Hayling on Tuesday.

The bridge will have a span of 720 feet from tower to tower, and along the steel girders a traveller car will run from side to side. The dimensions of the car will be 50ft in length and about 30 or 35ft. wide, and it will be capable of carrying an electric car and about three or four carriages, besides pedestrians and cyclists, up to about 60 tons in weight. While should it be required for ferrying troops across this stretch of Langstone Harbour, up 5,000 could be conveyed from one side to the other in the course of an hour.

The illustration shows the level of the water at low ordinary spring tide, when the distance from the girder would be 103ft. 6 ins., and at high tide when the distance would be 90ft. The lower part of the car would be a distance of 15ft. above the high water level, and 28ft. above low water. The elevation of the bridge it is estimated would be quite high enough to allow merchant shipping passing underneath.

Mr G.E. Eady M.I.C.E of 53 Victoria Street, Westminster and Mr Cobbett M.I.C.E. of Fareham, are the engineers of the scheme, and are confident of the ultimate success of the undertaking. Similar bridges have been or are being constructed over the Usk at Newport (Mon), across the Mersey at Runcorn, and over the Ribble between Southport and Lytham, whilst others are at Rouen over the Seine and at Bilbao, Spain, have given every satisfaction. Everything is to be worked by electrical power.

Several gave evidence as to the necessity for the bridge and the great advantage of linking Hayling and Portsmouth. The cost was estimated at about £63,000. The bridge would be built sufficiently strong to be safe in a gale of 80 miles an hour without any stress and it would stand in a storm up to 160 miles an hour wind velocity.

The engineer was not 'G.E. Eady' as stated in the newspaper report, but the eminent Westminster-based engineer George Griffin Eady, who was also named, along with Alfred S. Frech, as one of the proposed engineers for two other bridges which were never to be built – across Poole Harbour in Dorset linking Sandbanks with the Isle of Purbeck, and the 1905 proposal for a bridge across the Tees where the 1911 Middlesbrough bridge now stands.

The Poole proposal was contained within plans announced by the Branksome Park & Swanage Light Railway Company, which had been established in 1904 to build a 13-mile tramway between Bournemouth and Swanage, crossing the mouth of Poole Harbour with a transporter bridge which would 'by means of a cage and chain arrangement … swing the cars across to the opposite bank and thence continue the system to Swanage.'

That intended bridge would thus have combined features of a transporter bridge with the well-established principles of the chain ferry – and would have been very similar, mechanically, to ideas contained within Ferdinand Arnodin and Alberto Palacio's American Patent No.425,724 for transporter bridges, granted on 15 April 1890. Arnodin would almost certainly have been consulted in the Poole bridge's design.

The intended drive system would have used electric motors under the gondola to draw the platform across the harbour mouth by gripping a tethered chain picked up from the seabed just as the chain ferry still does on the crossing today.

While the weight of the chain beneath the moving gondola would have aided lateral stability as the gondola traversed the harbour mouth, it would also have caused considerable drag, thus increasing the amount of power needed to drive it.

Both the tramway and bridge proposals were subjected to an enquiry by the Light Railway Commissioners who received numerous local objections to the plan – including from the Poole Harbour Commissioners and Poole Council and, in their report published in late March 1906, the idea was rejected.

Almost a quarter of a century later, legislation passed by the Dublin Government – The Dublin Port and Docks (Bridges) Act of 1929 – included provision for two new bridges across the Liffey, one of which was for a replacement for the centuries-old Butts Bridge. The other was to be a transporter bridge with a much higher capacity than any other transporter

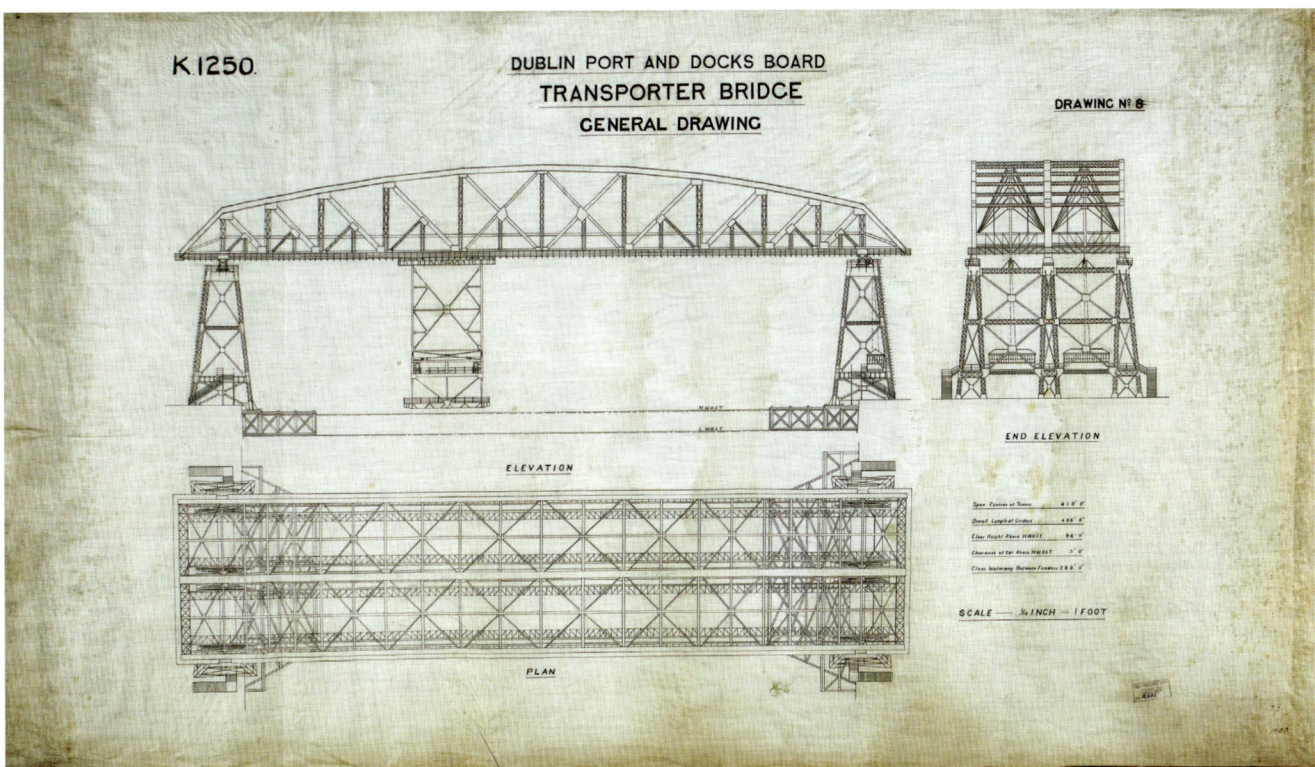

ever built. This gave the Port Authority wide-ranging powers – assuming the bridge would be completed within seven years of the passing of the Act. The Act empowered them:

> 'To construct and maintain a new high level Transporter Bridge over the River Liffey with all convenient and necessary abutments piers hydraulic or electric machinery and works in connection therewith and all necessary approaches thereto for the passage and conveyance of passengers horses cattle carts carriages and vehicles of every description, such bridge commencing from a point near the southern end of Guild Street on the North Wall Quay and terminating at a point near the northern end of Cardiff Lane on Sir John Rogerson's Quay.'

Had the bridge – proposed by Dublin Docklands Engineer Joseph Mallagh – ever been built, it would have had a span of 120 metres and have crossed the Liffey at the same point as Santiago Calatrava's 2009 Samuel Beckett Bridge does today.

The unique design envisaged very wide towers to accommodate the docks for the twin gondolas which would be able to pass side by side in mid-river – each gondola having a carrying capacity of 16 vehicles on the lower deck and up to 700 pedestrians on the upper. With both operating, there would be a crossing every three minutes. The main beam would have been under immense stress when the two fully laden gondolas passed mid-stream, hence its robust design.

The scale drawing of Joseph Mallagh's proposed Dublin Transporter Bridge across the Liffey is preserved in the archives of the Dublin Port Authority. The bridge bears a marked similarity to the Puente Nicolás Avellaneda in Buenos Aires. An illustration of a wooden model of the proposed bridge appeared in the *Irish Independent* on 19 November 1928. The caption to that picture estimated the construction costs at £167,000. Crossing time for each gondola was estimated at three minutes, including loading and unloading.

The small gondola slung beneath the Maarsserbrug Bridge, the only transport bridge ever built in Holland.

A two-gondola design had only ever been proposed once before – by Ferdinand Arnodin for his planned bridge over the Garonne at Bordeaux. That project – which was never built – is discussed in depth in the following chapters.

The seven years passed, followed by a three-year extension of the Port Authority's construction powers, with the plans gathering dust. Then followed what was known as 'The Emergency' in Ireland – the Second World War – and while the bridge idea was briefly revived in 1948, it was soon to be abandoned.

The last transporter bridge to be opened had a short life. It combined a high level road bridge with a low level transporter gondola. That was the Maarsserbrug bridge connecting Maarssen-Dorp with Maarssenbroek in Holland. Built in 1938, a low-level transporter was suspended beneath the roadway to enable local farmers to transport their produce across the Amsterdam–Rijnkanaal, then known as the Merwedekanaal, without the lengthy detour which would have been needed to get to the high-level bridge's access roads. The gondola, little used since the end of the Second World War as farmers used Marshall Plan money to replace horses and carts with American tractors, was removed in 1959.

Despite their obvious cost-effectiveness in construction, and functionality in use, transporter bridges had several inherent weaknesses in addition to strictly limited carrying capacity. One of the major disadvantages was that cable-slung gondolas were unable to work in very high winds, and the wider the crossing the greater the likelihood of wind disturbance.

In addition, with the gondola being slung so low that it crossed the river just above the normal high water level, any extreme rise in river level could also be a problem. For many transporter bridges, with the normal clearance of the gondola deck somewhere between 3 and 5m, in periods of extreme rainfall, there could be major issues.

Nowhere was that more apparent than at Rouen where the Seine periodically rose to such a level that it flooded the quayside and submerged the gondola. In both 1910 and 1924 the river rose more than 6m above its normal high water level, and took the bridge out of service for weeks at a time.

Further upstream, at Paris, the river level had risen more than 8m on both occasions, and as a visual reference, that was the same sort of flood levels which inundated Paris and several other locations along the Seine in the spring of 2016.

Not only did that cause the gondola to be submerged, it also exposed it to a much greater risk of damage as fast-moving floodwater brought debris downstream with it.

But it was their limited vehicle capacity at a time when road transport was increasing exponentially which brought the relatively brief era of the transporter bridge to an end. Some struggled on into the 1930s but one by one they were supplanted by fixed high level bridges and demolished.

When the River Seine flooded in February 1924, the gondola on the Rouen bridge was submerged for some days.

C. V. 27. ROUEN (30 Janvier 1910). — La Crue de la Seine
Le Transbordeur et le Quai du Havre.

With the gondola normally travelling across the river just a few metres above high water level, in times of flooding its submersion was inevitable. These two postcard views show the effect of the River Seine flooding in 1910.

C. V. - 33 - ROUEN. - 31 Janvier 1910 - La Crue de la Seine
La Nacelle du Transbordeur, rive gauche

A few lasted a lot longer simply because the lower volumes of traffic using them made their replacement by conventional high level bridges uneconomic. They were the lucky ones, for they survived into the modern era where the preservation of such important structures as visitor attractions for the growing tourist industry was both actively encouraged and considered economically sustainable.

Today, nobody would even consider demolishing such remarkable structures if there was chance of refurbishing them – the rebuilding of the Rochefort bridge is a case in point.

The majority of the transporter bridges which were actually built have, however, been lost over the years, including all but two of those designed by Ferdinand Arnodin. Most of those became casualties of the Second World War, having all previously – and almost miraculously – managed to survive the ravages of the First World War.

The only survival of Arnodin's bridges in France, the 140-metre Rochefort–Martrou Bridge in Charente-Maritime, while normally operational in summer, is currently closed until 2020 for major restoration work, and the Newport bridge over the Usk – the only one of Arnodin's designs to be built outside his native France – is now restored after a period of closure and still operates during the summer months. The restoration of the Rochefort bridge – effectively a re-build of everything except the towers – is illustrated in a later chapter, while the history of the Newport bridge is also the subject of a separate chapter later in this book.

All of his other 'ponts transbordeur' have been either destroyed or demolished – at Brest, Rouen, Nantes and Marseille, as well as the never-completed bridge at Bordeaux.

Amongst the others which have disappeared – the work of several other designers and engineers – are the bridge over the entrance to the naval harbour at Kiel, Germany, two of the three built in Buenos Aires, one in Rio de Janiero, and two in England: the Widnes–Runcorn bridge and Crosfield's No.1 transporter across the Mersey at Warrington.

Although it is now more than a century since the last transporter bridge was constructed – Joseph Crosfield's second bridge over the Mersey at Bank Quay in Warrington was completed in 1915 – their story is far from dead.

Today, the ever-growing importance of the tourist market has led to the idea of transporter bridges being revisited for leisure rather than commercial traffic – the role of the bridge changing from being a means of getting across an otherwise impassable divide to being a destination in its own right. Over the past few years several new projects have been mooted, although none has yet progressed beyond the feasibility study stage.

When the Royal Victoria Dock high level footbridge was planned for London in the 1990s, a second phase in its development envisaged a low-level glass transporter gondola travelling below it.

The bridge was opened in 1998 with the transporter docking quays in place, but more than twenty years later there is still no sign of the gondola being added.

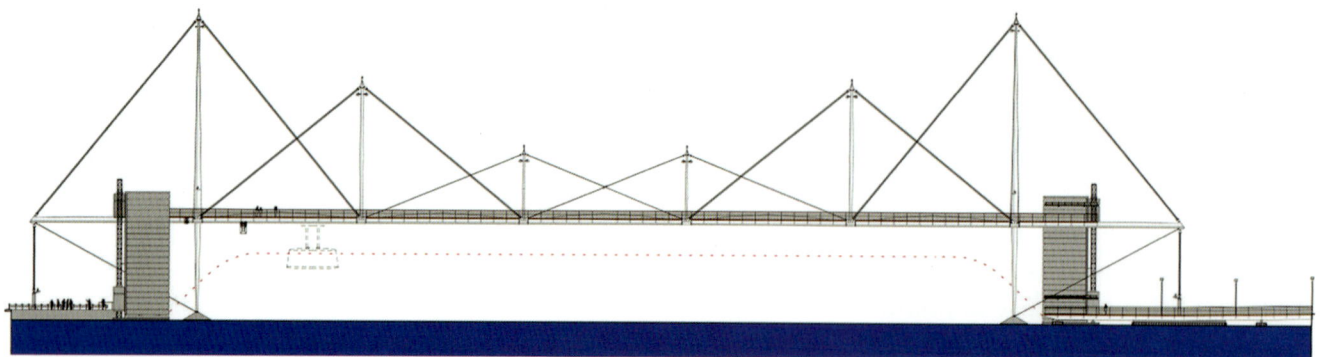

Above and right: The drawings by the architects Lifschutz Davidson Sandilands, showing how the gondola on the Royal Victoria Dock Bridge was intended to traverse the dock.

Far right: The docking station for the gondola remains unused as the transporter was never added to London's Royal Victoria Dock Bridge.

SECTION LOOKING SOUTH

ROYAL VICTORIA DOCK BRIDGE · Lifschutz Davidson

It would have been a unique take on the traditional design of the transporter bridge – the 4m x 6m glass gondola would have docked at low level, but then it would have been raised up as it progressed away from its docking station, crossing the water at high level before being lowered back down to just above water level at the opposite docking station.

Other proposals to built new transporter bridges, while employing modern materials and stunning designs, have followed the more conventional system of a low-slung gondola.

In Nantes in France 'Project Jules Verne' was conceived as a means of revitalising a somewhat run-down area of the port on the Loire. A cable-stayed 'pont transbordeur' on the Madeleine tributary of the Loire is seen as one way of bringing tourists to the area.

The 1905 Marseille bridge was featured in an American-published set of stereocards – 3D photographs – in the year of its opening, 1905, giving a clear view of its unusual cantilevered cable-stayed design.

NPPi - Paul Poirier architecte - 10-07-2013

Left and below: In these computer-generated images, a modern take on the transporter bridge is proposed for Marseille by Paul Poirier. This bridge would, if built, cross the harbour at approximately the same point as Ferdinand Arnodin's 1905 bridge. Like the original, Poirier's proposal is for a cable-stayed design, but with a rigid over-reaching stiffening beam carrying covered pedestrian walkways all the way across and linked to aerial cableways from other points in the city. Where the earlier bridge used vertical anchorage cables, Poirier's design would run anchor cables from the ends of the main beam down and inwards towards the bases of the towers.

The elegant and beautifully engineered ironwork of the world's first transporter bridge, the Puente Viscaya in Portugalete, Bilbao, Spain, designed by Alberto Palacio and engineered by Ferdinand Arnodin.

A local poll in 2012 voted strongly in favour of its construction, and there was once talk of it being completed by 2015. At the time of writing, work on the project has not yet started, and the future remains uncertain.

Behind the project is architect Paul Poirier, and a 'pont transbordeur' association has been set up to promote the idea of future transporter bridges. Another Poirier proposal is being promoted in Marseille to replace Arnodin's 1905 'transbordeur' demolished in 1940.

That £100M project has been met with a much more positive civic response. Despite little evidence of progress in recent years, the project seems to be gathering support.

In both of Poirier's designs, there are strong echoes of Ferdinand Arnodin's original concepts for the transporter bridges at Nantes, Marseille, Rochefort and Newport.

In Brest in 2013, Alain Masson, Vice-President of Brest Métropole Océane, published a proposition in the *Ouest-France* newspaper for a new bridge suggesting, 'we are heading towards a transporter bridge, reserved exclusively for pedestrians, cyclists and rollerbladers.' His proposal was for a cantilevered suspension design – a modern take on Arnodin's bridges at Nantes and Marseille.

Whether or not such bold and expensive plans ever bear fruit remains to be seen. If built, any one of these twenty-first century transporter bridges would be spectacular tourist attractions. The costs involved would make the €22.5M currently being spent on Arnodin's 118-year-old Rochefort bridge pale into insignificance.

The project budget for the Royal Victoria Dock Bridge in London was £5 million back in the 1990s – and the high-level bridge part of that project spans just 127 metres, compared with 109m at Brest, 141m at Nantes and 165m at Marseille.

Left and below: Two artist's impressions of a design for a new transporter bridge across the Madelaine branch of the Loire at Nantes, as imagined by French architect Paul Poirier and proposed as part of the 'Project Jules Verne' urban regeneration of the city's riverside areas. Taking full advantage of the greater strength of modern materials, this bridge, despite its elegant and minimalist design, has been projected to be able to transport a significant total load across the river, including a 'bendy-bus' alongside several cars.

As time goes on, and cost estimates rise, the likelihood of these projects ever being built seems less and less likely, but, if they were constructed, what popular tourist attractions they would become.

Perhaps such bold visions are exactly what is needed to trigger the redevelopment of abandoned industrial areas.

THE WORLD'S FIRST TRANSPORTER BRIDGE

From an early age, Don Martin Alberto Palacio y Elissague was fascinated by the architectural potential of cast iron, steel and glass. Born in 1856 in Northern Spain, he studied architecture, structural engineering, mathematics and medicine in Paris, and one of those who participated in his education was Gustave Eiffel, whose influence would have a profound and enduring effect on the future direction of Palacio's career.

Back in his native Spain, Palacio was one of the central figures in the development of Madrid's Retiro Park, collaborating as the structural engineer with the architect Ricardo Velázquez Bosco in the construction of Palacio de la Minería for the 1883 Exposición Nacional de Minería, the national mining

Opposite: Palacio's Puente Viscaya in Portugalete – the world's first transporter bridge – opened in 1893.

Above left: The Palacio de Cristal in Madrid's Retiro Park was an early example of Palacio's love of cast iron. Designed by Palacio and Ricardo Velázquez Bosco, it was completed in 1887.

Above right: Madrid's old Atocha Station was designed by Palacio and built in 1892. Since the adjacent new station was opened, it has been given a new lease of life as an exhibition centre and sub-tropical plant house.

The bridge, photographed in 2017, operates around the clock, as the alternative means of crossing the river Nervíon requires a 24-kilometre round trip.

exhibition. The building, now an arts centre, is known today as the Palacio de Velázquez and is the only survivor of several pavilions erected for the exhibition.

He collaborated with Bosco on another project in the Retiro, this time in creating Madrid's homage to London's Crystal Palace – the Palacio de Cristal, which was completed in 1887.

His importance in this book, however, derives from his work in Bilbao between 1887 and 1893, designing and constructing the world's first transporter bridge – originally to be known as the 'Puente Transbordador Palacio', but this was later changed to the 'Puente Colgante'.

Today it is more usually referred to as the 'Puente Bizkaia' or 'Vizcaya Bridge', linking the towns of Portugalete and Getxo, or Las Arenas, across the mouth of the Nervíon River.

In that project Palacio as the project's designer engaged the French engineer, Ferdinand-Joseph Arnodin, whom he placed in charge of construction.

The relationship between the two men seems initially to have been an almost equal partnership, despite the fact that the bridge design was Palacio's. Arnodin already had a great deal of experience in engineering metal bridges – especially suspension bridges, which Palacio's design essentially was.

They both clearly believed that the transporter bridge had a great future, for they jointly patented their design in several countries – Palacio initially filing an application in the Patent Office in Bilbao on 5 November 1887, at the same time time as Arnodin filed in Paris.

Left: The carriage from which the gondola is suspended has been changed – and improved – on numerous occasions. The most recent innovation uses an articulated carriage with built-in electric motors driving its wheels.

Below left: Access to the bridge on the Portugalete side is down a narrow street between tall buildings. This view is taken from just in front of the Portugalete cable anchors.

Below right: The bridge dominates the town of Portugalete.

Just six weeks later, on 21 December 1887, they filed a further application in London. Their specification was accepted and British Patent No.17,573 was issued in their names on 27 January 1888.

Like many patents at the time, it was a 'cover-all' description of a transporter bridge, for none had yet been built. Their patent covered many possible design solutions to the challenge, and bore similarities to Charles Smith's description of his proposed bridge published in *Engineering* on 25 July 1873.

Indeed, the British Patent specification enumerated several different design types, and could, in effect, have been used to cover the development of all five of Britain's transporters as the designs of all of them were effectively covered by Palacio and Arnodin's patent rights.

The two men also registered their design in Belgium – on 29 February 1888 – and submitted a patent application to the United States Patent Office on 26 March 1888 which was given Serial No. 268,526.

With the idea protected in most of the countries where they foresaw that such bridges might be built, their commitment to the idea seemed robust.

There was, as might be expected, a great deal of duplication in the wording of the various patents, but there were also significant differences. In their British patent, they suggested several different motive power sources for the gondola carriage – including steam, compressed air, vacuum and electricity. The term 'gondola' does not appear in the British specification, Palacio and Arnodin's UK agents, Jenson & Sons, preferring the term 'transporting basket'. The specification also suggested a much more robust structure than was used in any of Arnodin's subsequent bridges.

'The transporting basket serves at once for a ferry boat which acts in the air and has not to contend with the inconveniences from the influences of currents, tides fogs or waves and other obstacles. Being always at the same height and following a regular line in its progress it does not present the difficulties of embarkation and the drawbacks that arise from the variations in the elevation of the ferry boats caused by the ebb and flow of the tides; and it will work at night time with the same facility as it will at day time.

The rods which carry the basket are attached triangularly and hereby withstand the troublesome swinging produced by the wind in the passage.'

It appears that they intended to use a rigid suspension frame for the gondola rather than the cables which became the normal practice.

The American patent just weeks later suggesting cables or rods might be used, and the 'suspending basket' had become the more robust and safer-sounding 'platform or car'.

'Our invention relates to means employed for transporting a load from one point to another across a water-course or ravine, or where the conditions are not favorable for making a road, or where it is impossible or difficult to build an ordinary bridge.

The Portugalete shore on the Puente Viscaya's opening day, 28 July 1893. This image was published as a postcard around 1903. Perhaps the storm which had swept in from the Bay of Biscay that day kept onlookers away.

The bridge in 1939, rebuilt after the Civil War, with the gondola approaching Portugalete. The covers over the cable anchors were to stop condensation rolling down the cables and dripping into the passenger cabin.

In carrying out our invention we employ a girder or light bridge-like structure supported on piers, towers, or other elevated supports at a sufficient height – say high enough for the highest vessel to pass under and provided with rails or ways, upon which runs a frame-work mounted on rollers, to which are attached metal rods or cables for suspending a platform or car, which may be, for example, at the height of the banks or quays of a river or water-course, and within which or on which are carried the passengers, railway cars, or other vehicles, animals, or other merchandise which it is desired to transport across the river, water-course, ravine, or space which is spanned by the girder.'

Right: A postcard view of the Puente Viscaya, taken from Portugalete around 1910. The night-time photograph shown earlier is taken from approximately the same viewpoint and shows the extent of subsequent urban development in Las Arenas.

Below left: Looking towards Las Arenas, in this postcard from around 1904 the electric winch house can be seen above the landing stage between the two towers.

Below right: From Las Arenas, the rear of the electric winch house can be seen, with the lift shaft to the right of it within the right-hand tower.

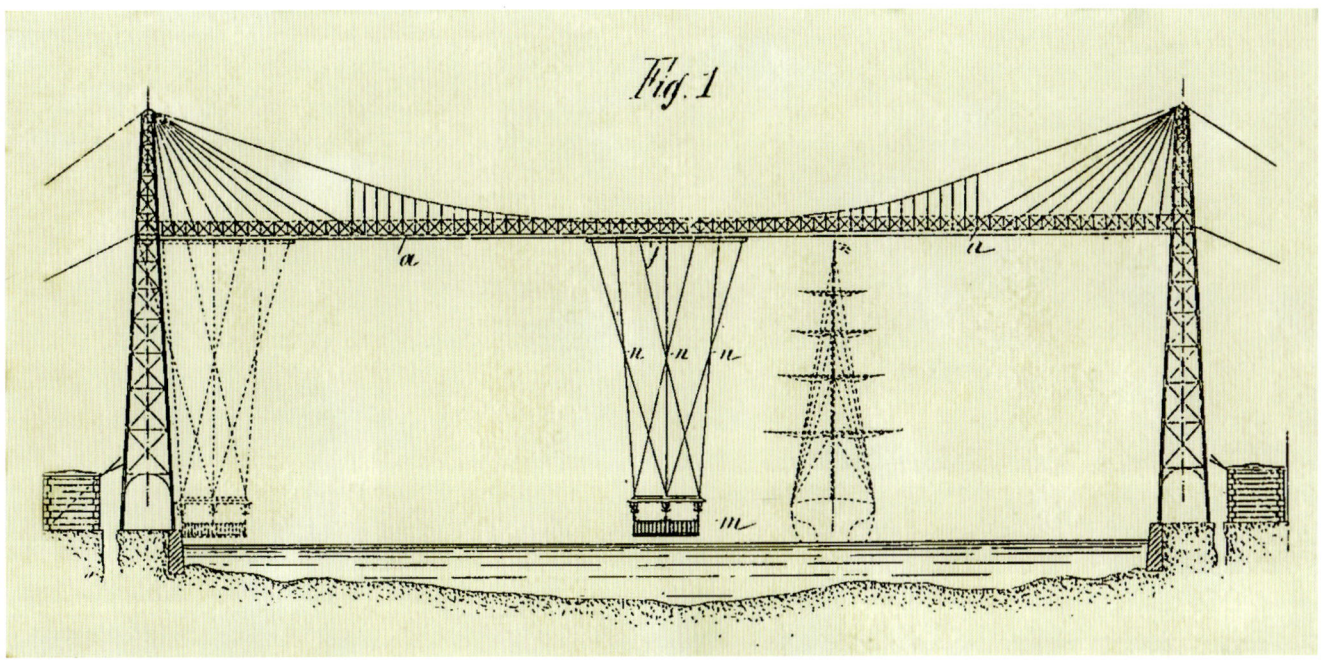

Although the same in design principles, the bridge which Palacio and Arnodin actually built near Bilbao looked very different to Smith's 1873 proposals and worked in a very different way to their own American patent specification in which they had suggested a propulsion system based not on cable winches, but on the established principle of the chain ferry.

'We will now describe how the movement is given to the transporting car or platform and the rolling frame from which it is suspended. Across the river or water-course therein represented there is laid from one bank to the other a cable or chain, the ends of which are firmly secured to the shore. This chain or cable is long enough to lie upon the bottom of the river with considerable slack. This cable or chain passes under the car or platform and over a chain wheel or pulley, the shaft or axle of which is supported in bearings fixed under the car or platform, and the said chain wheel or pulley has rotary motion given to it in either direction from a reversible motor, placed upon the

Palacio and Arnodin's original 1887 patent design for a Transporter Bridge, combined vertical suspension cables with angled cable-stays.

The original 1892 drawings for Palacio's Puente Movible or 'hanging bridge'. Arnodin's influence may have been behind the decision to use French for the left-hand panel.

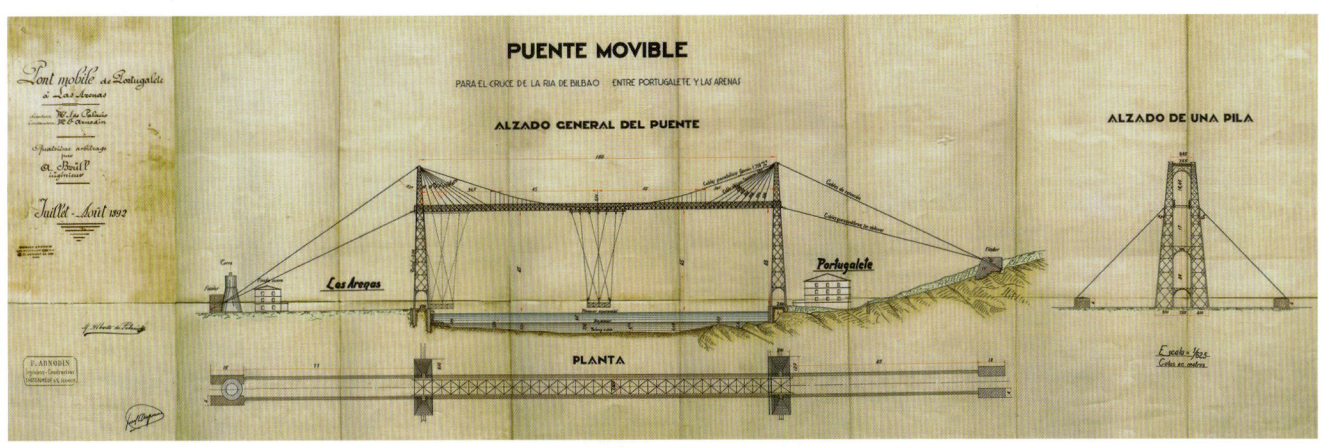

Above: The view along the quayside at Portugalete.

Inset: Detail of the drive gearing, from Palacio and Arnodin's patent specification.

floor of the car or platform. This motor may be of any convenient or suitable kinds, for instance, a steam engine or an electric motor… In passing over the pulley or wheel the chain is lifted from the bottom to plunge again into the Water as soon as it passes, thus leaving the navigable channel completely free. This means of propelling the car or platform will prevent any oscillation of the suspended car or platform by taking hold upon the bottom of the river or watercourse. Moreover, this method of propelling or giving motion to the car or platform insures it always arriving at the desired landing-place, and being there securely held during the embarkation and disembarkation.'

Right: Palacio and Arnodin's original 1887 patent design for a Transporter Bridge combined vertical suspension cables with angled cable-stays.

Far right: One of the 12 electric motors which drove the carriage.

As far as research can ascertain, no transporter bridge was ever constructed using the anchored-chain system, and it is interesting to note that, with work on the Viscaya bridge already underway when this patent was filed, Palacio planned to move the bogie car across the beam using a compressed air system invented by Louis Mékarski in the 1870s and used to pull trams in Nantes and, for a time, on the Caledonian Road tram system in London as well.

In the end, he used steam winches powered by a two-cylinder vertical steam engine built by Henri David in Orleans and set high in the tower on the Las Arenas side of the river. An elevator was installed from the outset, not initially for the benefit of passengers, but simply to raise the coal and water – drawn from the nearby River Gobelas – up to the winch engine. From the bridge's earliest days local people did, however, pay a small fee to be allowed up on to the beam.

When electricity came along in 1905, the steam winch was replaced by an electric winch. Steam's role thereafter was powering an electric generating station. Electricity was also used to power the elevator within the Las Arenas tower. Later on, the electric motors were augmented by a petrol engine as the mains electricity supply was notoriously unreliable.

Below left: As the carriage passes along rails on the underside of the beam, the electric cabling and direct-drive motors can be seen. Originally, there were two slender walkways, one at either edge of the beam. Today there is a single central aisle walkway.

Below right: Looking through the steelwork from about half way along the beam as the gondola sets off from Las Arenas.

A view of the construction site with the towers nearing completion, taken from near the Basílica Santa María de Portugalete, probably in the autumn of 1891.

The illustrations in the patent specifications suggest that the principles of Palacio and Arnodin's ideas had been fully worked out long before work on the Bilbao bridge started.

The combination of suspension cables supporting the central part of the main beam and cable stays taking the strain near the towers is clearly illustrated in their 1887 drawings. Stabilising a structure like this was crucial, and in the case of the Viscaya bridge, the load was initially shared by eight heavy cables on each side, anchored deep in the ground 110 metres inland from either tower. On the Portugalete side, the anchor stands on elevated ground while, in Las Arenas, the anchor is now surrounded by tall buildings.

Further lateral bracing came from cables running out about 60 metres each side at right angles to the bridge and anchored in the quayside parallel to the river to counter the effect of the strong winds which blow in from the estuary.

That basic design – but without the side stabilisers – would later be used by Arnodin in four of his bridges – at Bizerte/Brest, Rouen, Rochefort-sur-Mer and Newport. Today, however, the Viscaya bridge's main beam is entirely supported by suspension cables – one of several major modifications dating from the rebuild after the Spanish Civil War.

The only other transporter bridges to employ side bracing were Arnodin's Bizerte bridge and the Crosfield No.1 bridge at Warrington designed by James Newell and built by Thomas Piggott in 1908.

Arnodin had pioneered the use of multi-strand wound cables for his earlier bridge designs, and the bunches of cables were bound in canvas and tar to

make them weatherproof. Since rebuilding, the number of tensioning cables has been reduced to four behind each tower.

The ironwork for the towers was forged in Le Creusot and mild steel cables came from the Firminy wireworks. Arnodin's factory in Chateauneuf-sur-Loire was responsible for putting the whole structure together, first in the factory then on site.

Palacio was already a well-known architect, with several notable buildings to his name, when he began work designing the transporter bridge. He would doubtless already have been well aware of the many proposals for 'hanging bridges' which had been published over the preceding decades, but his unique refinement of the idea would be the first to be built.

As the population of the area grew and prospered, there was a pressing need to develop the lands on the opposite side of the river from Portugalete and to provide a more reliable and safer method than that which was then offered by small ferry boats – and to do so without interfering with the growing river traffic which accessed the expanding port of Bilbao further upstream.

In the late 1880s, the port of Bilbao was much closer to the city, and the River Nervión was the main entry point for the goods which kept the city alive, and the main exit point for its produce and heavy industry. On some days there could be dozens of tall ships navigating the narrow channel, so a bridge which

The bridge after the main beam had been demolished towards the end of the Spanish Civil War. The towers remained upright. The wreckage was quickly cleared from the navigable channel to allow supplies once again to reach Bilbao and, in 1939, government approval was given for the reconstruction of the bridge as quickly as possible. By 1941 a new beam was in place and the bridge was open for traffic once again.

Above left and right: The vertical suspension cables are each clamped to a pair of main cables.

gave them primacy was of great importance. Docks and quays lined the river almost all the way from Bilbao's city centre to just upstream from the site of the bridge. Networks of railway lines delivered and received goods from all over Northern Spain. Palacio's solution probably owed much to earlier – and never realised – designs for similar bridges in Britain and America, but his was simpler, more elegant, and above all practicable.

Design work on the bridge started in 1887 with the construction of the stone piers on which the towers, or pylons, would be erected, starting in 1887.

Construction time was three years and the project was underwritten by Santos López de Letona who became principal shareholder of the bridge company.

The walkway across the beam of the Viscaya bridge has lost some of its original open elegance in meeting the needs of today's health and safety requirements.

The Portugalete tower reflected in a modern building – the bridge is the tallest structure in Portugalete and dominates the area.

The site chosen was not particularly narrow – a span of 160 metres – but Palacio's ability to design cast iron structures of immense strength with minimal metalwork resulted in a design which has withstood daily use now for 127 years. Of course, when opened in 1893, no-one could have envisaged that the bridge would one day carry anything approaching today's gondola loads. All it was intended to carry – initially on little more than a roofless open suspended platform – were local horses and carts, and commuting pedestrians.

A clearance of 45 metres below the main beam ensured that the tallest ships then known could pass safely beneath the bridge at all stages of the tide.

Irrespective of the actual design of the Viscaya bridge – transporter bridge design would evolve over the years – the original patents had all described a wide range of possible power systems, suggesting that once again Palacio and Arnodin were attempting to 'cover all bases' in their specifications – with an eye, perhaps, to the future enforcements of their patent rights.

> 'The movement may be effected by a rope actuated by a motor placed on the bank and of a construction suited for the circumstances such as a steam engine, a compressed air engine, or a hydraulic or electric motor. In case of using an electric or compressed air motor the wind may advantageously be utilized as motive power by means of a windmill or other air motor placed for instance on the top of the pillars of the bridge and accumulating the power in one or more accumulators sufficient for the intermittent working or passage of the basket and also for lighting by night by electricity if required. In case of a calm or insufficient wind an auxiliary motor may be applied.'

There was much original thinking behind such ideas, but much of the technology necessary to make it happen was still years in the future. Despite the fact that nobody had yet made the idea work, even to have thought of harnessing hydraulic power to drive and light their bridges was pioneering. Some of their ideas, however, would be developed into practical applications – albeit by others. For example, the patent specification made much of their advocacy for direct electric drive, despite the fact that the gondolas on the Viscaya bridge and several later transporters would be driven using steam-powered or electric-powered land-based winches.

> 'In case of using electricity the moving cord may be omitted and the motor made to form part of the rolling frame and the electricity may be conveyed thereto from the bank by suitable conductors; this motor would then set in motion a tooth wheel on the rolling frame in gear with a rack fixed on the bridge. This arrangement admits of passing the conducting wires by way of the basket by means of guide pulleys suitably arranged and the attendant in charge can thus freely manage the conductors for working the motor and the rolling frame in one direction or the other or stopping it.
>
> Or, as we prefer it and as assumed in the drawings the motor may be placed in the basket where it is most easy to look after it and control its working.'

Only the Duluth bridge in USA would briefly use this arrangement of housing the motors in the gondola itself, while the first to use DC motors directly driving the wheels on the carriage would be the Runcorn–Widnes bridge in Cheshire. That latter idea was abandoned on the Cheshire bridge very early

on it its life, and it has taken modern ingenuity and engineering to arrive at a system on the Viscaya bridge which works flawlessly.

Being the first in the world, many of the features of Palacio's bridge were innovative, and despite the stiffening beam being destroyed by explosives during the Spanish Civil War – and later rebuilt and extensively modernised in the last century and a quarter – the structure's appearance remains largely as he originally conceived it.

However, all that survives of the 1893 bridge are the pylons themselves, and one perhaps surprising fact – given that Palacio was a Spaniard and a patriotic Basque as well – is that the ironwork for the towers came from France, cast in French foundries, and not from the rich seams in the hills above Portugalete, a town which also boasted several very successful iron foundries equally capable of the casting work.

The superstructure was hot-riveted together on site – a total of more than 400,000 rivets – the assembly taking nearly three years.

Support for the latticed beam combined the already proven technology of vertical suspension cables, together with angled cables running from the tops of the towers which we are today familiar with on cable-stayed bridges. As mentioned, the rebuilt steel beam is suspended from vertical suspension cables alone – 70 of them.

Twenty years ago, the drive was changed yet again, to the current system of 12 electric motors providing direct drive on to the bogies of the travelling

The bridge operates around the clock – although the fares rise out of normal working hours. This photograph was taken from near the Basílica Santa María de Portugalete in the heart of the medieval old town.

The lateral cables which stabilise the bridge are embedded into the quayside on either side of each tower – a precaution deemed necessary because of the gusting winds which blow in from the estuary. This is the quayside on the Portugalete side of the river.

The *Scientific American* cover with Palacio's Columbus Monument design.

frame. This cleared the stiffening beam of winch cables, allowing its conversion to today's pedestrian gallery.

Surprisingly, given his considerable investment in establishing such wide-ranging patent rights, Palacio never designed another transporter bridge, but he had already suggested some projects which would have been on an even grander scale. The cover of the October 1890 issue of *Scientific American* magazine illustrated of the grandest of them all – a massive six million dollar monument to commemorate the 400th anniversary of Christopher Columbus's discovery of the Americas. That estimated cost would equate with between $7bn and $8bn at today's prices. Needless to say, it was never built, but the *Scientific American* article described it in great detail.

'When North America proposed a competition for the construction of a tower to be erected at the Universal Exposition of 1892, a Spanish architect, a native of Bilbao, Mr. M. Alberto de Palacio, drew an original design, of which we publish an engraving. Mr. De Palacio has conceived a most perfect form, the sphere, which could not have been used prior to the knowledge of iron as a building material, because only by the modern methods of uniting the various parts, of which this material is susceptible, could a sphere be produced with a diameter of nearly 1,000 feet, that is, equal to the height of the Eiffel tower. This idea symbolizes the geographical completion of the earth which was realized by Christopher Columbus' discovery of the New World. The following is a description of the magnificent design: The colossal sphere is mounted on a base which is 262 feet high, and is crowned at its North Pole by the caravel which carried Columbus to the New World. The monument is brilliant with the colors of the continents, oceans and islands of the terrestrial sphere.'

Palacio had developed a business plan, estimating that:

'One hundred thousand spectators paying an entrance fee of $1, will bring $100,000. This will replace the capital in 62 days, without counting the profits of the cafes, entertainments, etc. The estimated total cost is $6,000,000.'

The article ended by noting that 'The architect, Mr. Palacio, is the designer of the movable bridge at Bilboa' – the idea of calling them 'transporter bridges' in any language had yet to be developed.

Monuments to Alberto Palacio still stand in small public spaces adjacent to each of the Puente Viscaya's towers – one of them unveiled only very recently – and given the success of their early partnership, it is surprising that he seems not to have continued his collaboration with Ferdinand Arnodin.

Had there been some sort of rift between the two men which brought such a close collaboration to an end? If so, an account of what might have caused it has yet to be discovered.

Having put his name to patents across the world – and presumably having met his share of the patent application costs – Palacio appears to have shown no further interest in the development of transporter bridges.

Far left: On the Portugalete side, a modern bust of Palacio sits below the cable anchors and looks out over his creation.

Left: On the Las Arenas, or Getxo, side of the river, a bust was erected in 1956 to mark the centenary of Palacio's birth.

A Canadian patent from February 1896 was registered in both names and follows the format of the London 1887 application, but later patents covering cantilevered and cable-stayed designs are in Arnodin's name alone.

The two men had commenced secret discussions about the design of the Portugalete bridge in the summer of 1887 and by the time they gave independent presentations of their proposals in Paris and Bilbao in November that year, their patent applications for the 'transbordeur' had been deposited simultaneously – at 11am on 5 November – in the Paris patent office by Arnodin and at the Bilbao office by Palacio.

So they were joint patent holders as well as apparently seriously committed business partners, but their active partnership seems to have lasted little more than a decade. Presumably Palacio still earned his share of income from the development of Arnodin's future bridges – although, again, there does not seem to be any extant documentation to that effect. By the late 1890s, his name had largely been erased from the story of the transporter bridge.

In his essay to mark the inauguration of the world's third transporter bridge, titled 'Le Pont a Transbordeur de Rouen', which was published in the *Précis Analytique des Travaux de l'Academie des Sciences, Belle-Lettres et Arts de Rouen* in 1900, M.A. de Pillon de Saint-Philibert attributed the transporter bridge solely to Monsieur Arnodin, 'who had invented the idea himself' – with no mention of Palacio's involvement whatsoever. The man who had conceived, designed and commissioned the first successful transporter bridge to be built in the modern world had been almost totally airbrushed out of its history by the more flambouyant and entrepreneurial Monsieur Arnodin – except in Palacio's native Basque country of course, where his celebrity remains unchallenged.

A bust of him was unveiled near the Las Arenas tower of the bridge on the centenary of his birth.

THE 'SYSTEME ARNODIN'

Ferdinand Arnodin's is the name most frequently associated with transporter bridges, or 'ponts transbordeurs,' yet only two examples of his transporters survive. After working with Palacio on the Puente Viscaya, he took on both the design and the construction of more bridges – several of which have an 'off the shelf' commonality of components, despite the fact that each new crossing posed unique challenges – and was a consultant on several others.

After the Bilbao bridge, the next two bridges to be completed were in Bizerte in Tunisia in 1898 and in Rouen the following year. Both are usually cited as the work of Arnodin on his own, although they bore many similarities to Palacio's design for the bridge in Bilbao. Both were shorter spans – the Rouen bridge at 143 metres and 109 metres for Bizerte, compared with the 164 metre span between Portugalete and Las Arenas.

As with the Puente Viscaya, drive for the Bizerte bridge was supplied by steam winches, a winch-house being built within one of the towers. It can be seen in the illustration from a supplement to *Le Petit Parisien* published on Sunday 22 November 1903.

Despite its short life, the bridge offered a popular vantage point from where to view naval activity in the French navy's main North African naval

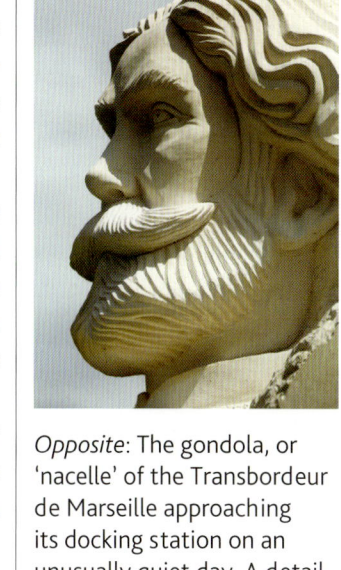

Opposite: The gondola, or 'nacelle' of the Transbordeur de Marseille approaching its docking station on an unusually quiet day. A detail from a Léon & Lévy postcard c.1904.

Above: A bust of Ferdinand Arnodin stands overlooking the Rochefort bridge.

Le Petit Parisien captioned this illustration 'Nos Sous-Marins – Le < Korrigan > franchit las Passe sous le Bac du Pont transbordeur' – literally 'The "Korrigan" passes beneath the deck of the ferry bridge'. The submarine *Korrigan* was launched at the Arsenal de Rochefort in February 1903. On completion she was assigned to Bizerte.

Right and opposite: The Brest transbordeur proved a popular subject for postcard publishers, and more than twenty different views of it have been traced, some of them tinted.

102 BREST. – *Le Port Militaire*. – *Le Pont Transbordeur*. – LL.

A 1903 view of the Bizerte bridge showing troops recently disembarked from the gondola. This view shows the eccentric positioning of the winch-house in the tower.

base, and *Le Petit Parisien* showed one of the country's early submarines – the *Korrigan* – passing beneath, to the obvious fascination of the gondola full of fez-wearing locals. A lot of artistic licence has been used by the originator of the illustration – the gondola is very much smaller and much simpler in design than that which was actually installed.

Like the Viscaya bridge, the load of the main beam was shared between vertical suspension cables and cable stays running from the top of each tower. The stability of the structure, as with any suspension bridge, depended on heavy cables embedded deep in the ground back from either tower.

As with the Viscaya, the Bizerte towers had side braces as well. Bizerte, or Bizerta, was one of several military harbours built for the French Mediterranean fleet, and the town itself was considered to be of major strategic importance.

The 'transbordeur' bridged the harbour mouth and became the subject of several locally published postcards. One such card, celebrating the visit of the French President Emile Loubet to Tunisia and Algeria in 1903, featured an image of the bridge together with the tricoleur and a portrait of Loubet.

111 BREST. — Le Port Militaire. — Le Pont Transbordeur. — LL.

88 BREST — Le Pont à Transbordeur

Three more postcard views of the Brest bridge showing its location within the busy 'Port Militaire'.

The bridge was dismantled after only nine years, in preparation for an expansion of the port and a widening of the harbour mouth to 200 metres – almost twice the bridge's 109 metres – in order to accommodate the navy's ever-larger vessels.

The transporter's component parts lay adjacent to the harbour for some months before being shipped to France and re-erected across the entrance to the inner harbour at the French navy's 'Porte Militaire' in Brest where it continued in use to transport people and munitions – on a very simple gondola compared with civilian bridges – until damaged by Allied bombing in 1944. It was demolished in 1947.

The transporter across the Seine at Rouen was very similar in design to the Bizerte bridge – although with a span 34 metres wider at 143 metres. Here, Arnodin tried out a completely new form of drive with two electric winches seated on a gantry above the gondola.

There were several other design developments, the most immediately noticeable being the more heavily cross-braced design of the towers themselves. It opened early in 1899 and survived until demolished in 1940 in the early stages of the Second World War, and perhaps because of the number of tourists the city of Rouen attracted, it seems to have been the most frequently photographed by postcard publishers. In the course of the research for this book, more than 60 postcards of the bridge have been identified, most of them published before 1920.

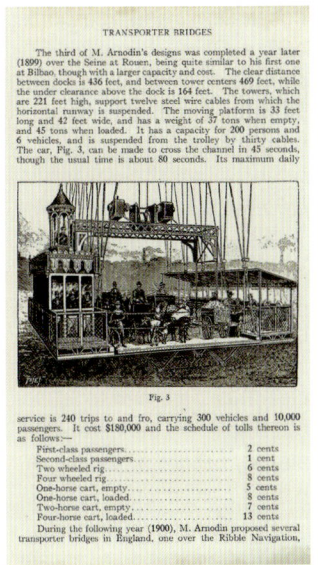

The gondola on the Rouen bridge, published in Henry Grattan Tyrrell's 16-page pamphlet *Transporter Bridges*, which was written in 1908 but not published until 1912.

152   ROUEN. — Le Pont Transbordeur. — LL.

The gondola on the Pont Transbordeur at Rouen had a unique design with the winches and cable reels mounted above the deck. The friction drive system was a typical Arnodin feature, whether on the gondola, as here, or in a winch-house. It was more resilient to damage than a toothed-drive should the bridge have to be stopped suddenly.

Rouen Docks and transporter bridge. Of all of France's transporter bridges, Le Pont Transbordeur de Rouen seems to have been the subject of most postcards, closely followed by Marseille. These views were published as postcards just a few years after the bridge had been opened – by which time, as steam replaced sail, the huge clearance designed to let tall ships pass beneath was already becoming less of an issue.

459 - ROUEN - Quai Gaston-Boulet

245. ROUEN — Le Pont Transbordeur, vu du quai du Havre

148. ROUEN. — Le Pont Transbordeur. — LL.

Arnodin seems to have been very willing to try out new ideas, stay with them if they were successful and abandon them if they were not. For the Rouen bridge, he introduced a very different form of cable drive, replacing the land-based motor house of earlier bridges with electric winches mounted on top of the gondola.

The system involved steel cables tethered at either side of the river, running to wheels on the overhead carriage, down to the winch reel on the gondola. While one cable was being wound on to the drum, the other was being unwound.

The system may not have worked as well as expected, for Arnodin never repeated it, although the designer of the Duluth bridge, Thomas McGilvray, having seen the Rouen bridge in operation, adapted the idea.

Arnodin's next undertaking, the Rochefort–Martrou crossing, showed a change in his thinking. While the basic design and construction of the towers was very similar to previous bridges – with cable-stays and vertical suspension cables supporting the 140 metre span, Arnodin evolved a very different profile for the steelwork of the towers, creating an elegant, sturdy structure which narrowed towards the base. He also employed an overhanging main beam with cable-stays either side of each tower, to spread the load more evenly as the travelling frame and gondola moved across the river.

It was clearly a design with which he was happy, for he used it, with necessary modifications of scale to take account of the different demands of each river crossing, on several subsequent bridges. The bridge was completed in 1900 and is now the only example of Arnodin's transporter bridges left standing in France.

Electricity to power the cable winches originally came from two steam engines and generators housed in a small power-house on the Rochefort shore, but in 1927, they were replaced by electric motors.

As with the Viscaya bridge, only the central section of the main beam was originally supported by suspension cables, but subsequent refurbishments and modifications over the years saw the inner cable stays replaced by suspension cables as well.

Detail from a postcard showing the unique gondola-mounted winch drive of the Rouen bridge.

Postcards of the Rochefort–Martrou bridge c.1905, including the photographer's horse-drawn wagon in the lower one. The building to the right of the lower postcard houses the anchor cables – modern photographs of it can be seen in the final chapter.

Édition spéciale de la Basilique

10. – MARSEILLE. – *La Nacelle du Pont à Transbordeur et N.-D. de la Garde*

NANTES. – Le Pont transbordeur
La nacelle au milieu du fleuve

Sous le règne de Sarradin
Maire de Nantes, Arnodin
Jeta par les soins de Baudin
Sur la Loire ce pont d'une Arche
Dont grâce à l'électricité
On voit avec docilité
La nacelle en notre cité
Du matin jusqu'au soir en marche.
D. CAILLÉ

The gondolas at Newport today, *above,* Marseille 1905, *top left,* Nantes 1903, *middle,* and Newport 1906, *bottom.* No safety gates in those days, just a chain.

Above: Today, on the Ile de Nantes side of the Madelaine – nearest the camera – the left-hand piers are beneath the present-day road bridge.

Far right: On the opposite shore, only the two piers built into the quay wall survive, the back piers now buried beneath the roadway below.

Right: Below the Ile de Nantes side of the 1975-built box girder Pont Anne-de-Bretagne, which bridges the Madelaine branch of the Loire in Nantes, stands one of the piers of Ferdinand Arnodin's 1903 pont transbordeur. One of the other piers has actually been incorporated into the structure of the road bridge itself. Several of Arnodin's transporter bridges were mounted on similar piers – at Rochefort (1900), Marseille (1905), and Newport (1906).

The almost deserted gondola mid-stream in the 1950s from the north tower looking down over the Madelaine towards the industrialised shore of the Ile de Nantes. The bridge was demolished just a few years after this photograph was taken. While the tower profile was very similar to Rochefort, the Nantes bridge – built five years later – had a cantilevered central section, the cable-stayed sections of the beam bearing most of the stresses.

At the time of writing, the bridge is nearing the end of a state-funded 3½-year programme to renovate it and restore it to its original configuration. That project is explored in the final chapter of this book.

In the early years of the twentieth century, as the demand for his bridges was increasing – bringing with it the need to establish more efficient production methods to reduce construction timescales – something approaching 'standardisation' in design was becoming evident. A tower profile similar to the Rochefort bridge was used in Nantes, built 1900–1903, and Newport, built 1903–1906. A variation of it was used for the cantilevered Marseille bridge in 1905. Despite their visual similarities, however, the total height of the towers varied from 66.5 metres at Rochefort to 87 metres at Marseille – and 92.5 metres for the never-completed bridge at Bordeaux.

There was also remarkable similarity between the gondolas at Rouen, Nantes, Marseille and Newport, so here again, perhaps, there is a sense of Arnodin developing some 'off the shelf' designs for some of the constituent parts of his later bridges. Contemporary postcards and photographs show just how close those similarities were.

1 - PONT A TRANSBORDEUR DE NANTES

Partie centrale en cours d'élévation. Poids : 46.000 kilog. à élever à 50 mètres au-dessus de l'eau

The Nantes bridge, during construction, the central cantilever section being raised into position.

The bridge shortly after completion.

Looking across the Madelaine from the Ile de Nantes as the gondola approaches – from a postcard c.1907.

210 - ROUEN - Le Pont Transbordeur. Ossature supérieure

The Newport gondola – the only surviving example of Arnodin's design – is still largely as it was built, more than 115 years ago. A few minor changes have been necessary to meet today's health and safety requirements, but otherwise Arnodin would probably be impressed and surprised by how well it has lasted.

One of Newport's attractions today is the opportunity to walk 'across the top'. Palacio's original idea of a pedestrian walkway across the main truss, which had been abandoned during the original construction of the Viscaya bridge, was included in the design for South Wales. A walkway had been incorporated into Arnodin's designs for the bridges at Rouen in 1899, Nantes in 1903 and Marseille in 1905 as well as at Newport. That the public was willing to pay 50 centimes – several times the cost of using the gondola – for the delight of climbing up steep stairs and then walking across the largely unprotected beam may have surprised the bridge builders, but it suggests that Palacio's business plan a quarter of a century earlier had been soundly researched.

Looking at the many postcards which were published of promenaders high above the ground, ladies dressed in long skirts, it is hard to imagine how they got up there – while some bridges had been fitted with elevators, several had not.

For those intrepid climbers who made their way up one tower and down the other, there were postcards available in the ticket office which could be stamped with the date the visitor had walked 'across the top'.

Above: This postcard was published five years after the Rouen bridge was opened. It clearly show Arnodin's ingenious cross-bracing on the sides of the main beam, an essential contributor to the beam's initial stability, but also its Achilles' Heel.

Below: A very early postcard of the Rouen bridge's beam.

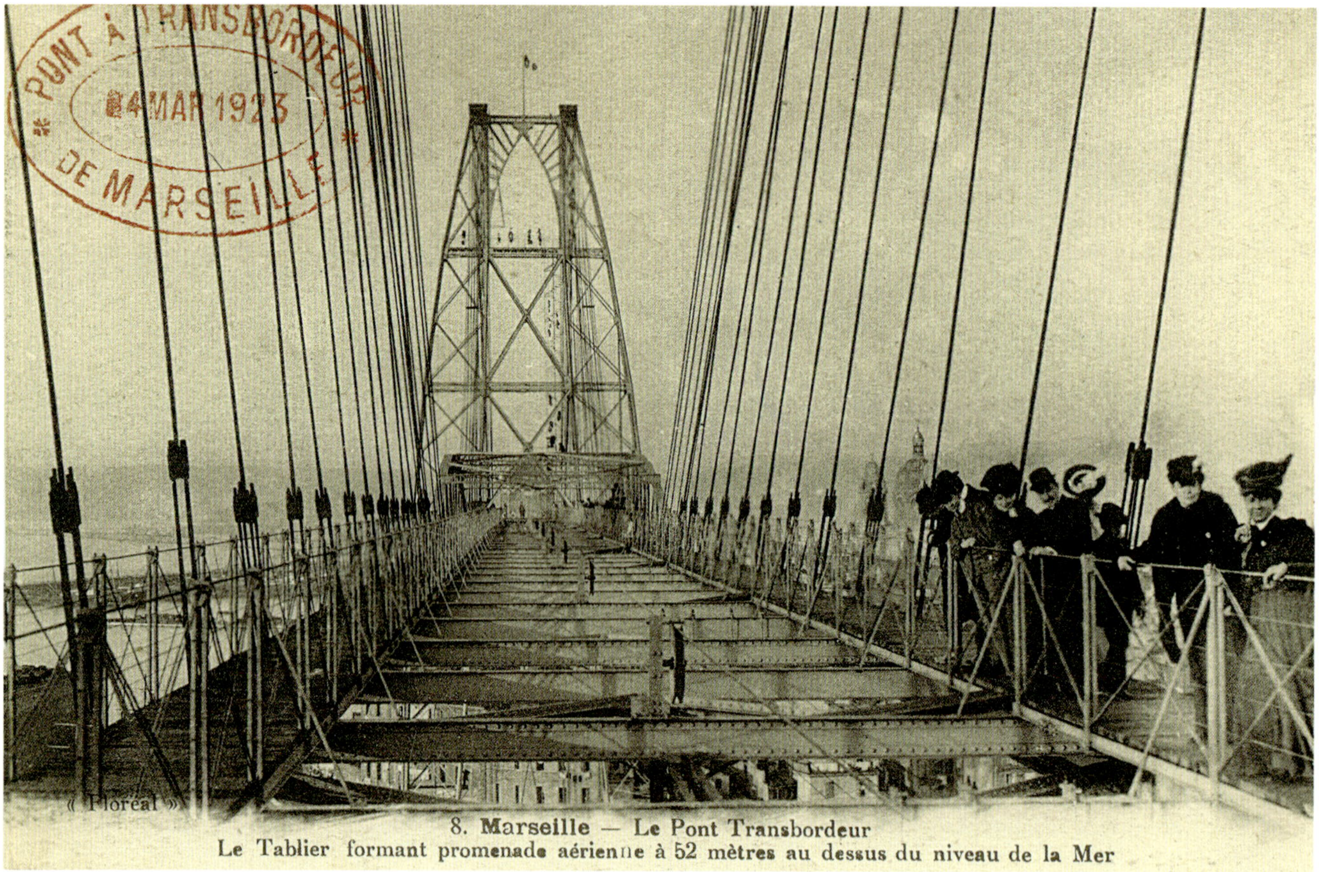
PONT À TRANSBORDEUR
24 MAR 1923
DE MARSEILLE
« Floréal »
8. Marseille — Le Pont Transbordeur
Le Tablier formant promenade aérienne à 52 mètres au dessus du niveau de la Mer

Above: A date-stamped postcard. 'Walking the top' became a popular pastime at several bridges – and at Marseille, there was even a buffet in the sky. Look closely, however, at the tower, where the intrepid few could climb the tower almost to the top.

Below left and right: Having a coffee and croissant in a location like this would have definitely required a good head for heights. Two of the many 1920s postcards which celebrated the bridge's unique attraction.

BUFFET
RESTAURANT

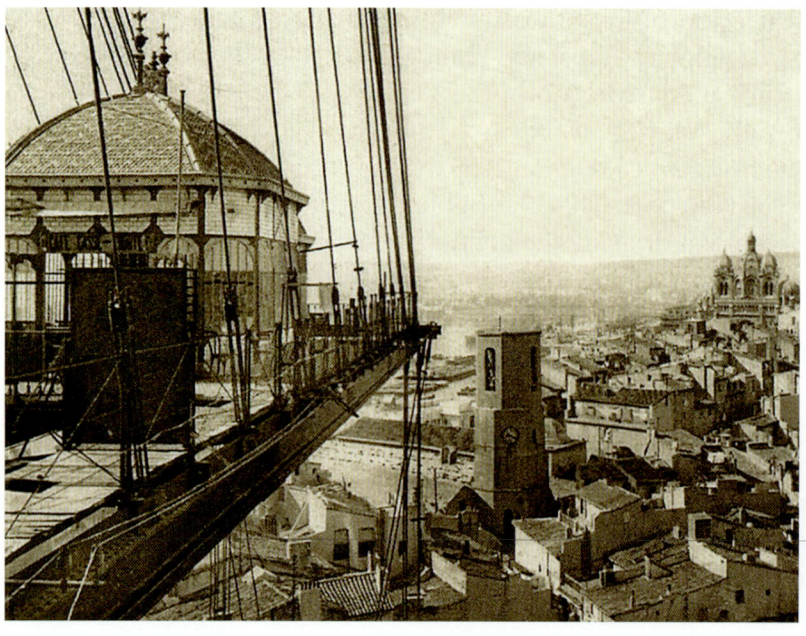

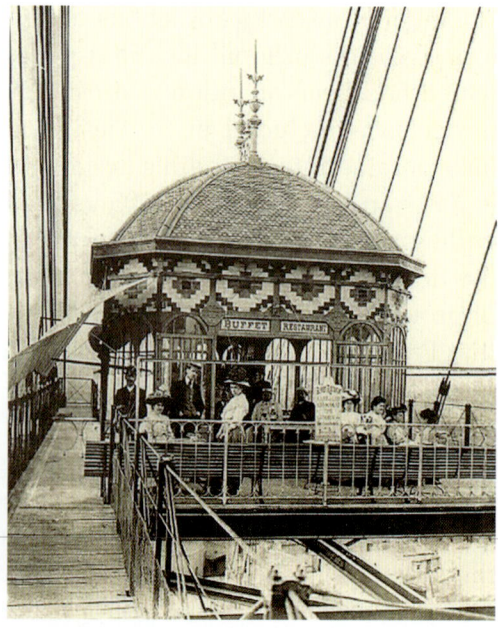

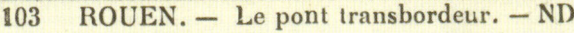

103   ROUEN. — Le pont transbordeur. — ND

Above: For 50 centimes, it was possible to climb several of Arnodin's bridges.

Left: An aerial view of the Marseille bridge in the 1920s.

Such cards either showed groups of other promenaders, or offered panoramic views of what could be seen from the top. To give an idea of relative costs, that 50 centimes was ten times the cost of posting any of these postcards in France at the time.

17. NANTES. — Nacelle du Transbordeur en marche.

Dating from the late 1920s, this view of the approaching gondola is taken looking towards the industrial skyline of the Ile de Nantes.

NANTES. - La plateforme du Transbordeur ou tablier avec les ascensionnistes qui viennent y admirer les superbes panoramas qu'on découvre de là-haut aux quatre coins de l'horizon.

Men 'walking the beam' in their best attire was one thing, but doing so in wide skirts must have presented the ladies with a unique challenge.

NANTES. - Panorama du Quai de la Fosse, près du haut du Pont transbordeur

Photographed from the walkway on the beam, the Nantes transporter casts a long shadow across the quayside.

208   MARSEILLE. — Le Pont Transbordeur. — LL.

Perhaps most surprising is the evidence from postcards of the Marseille bridge that pedestrian traffic 'across the top' was large enough to make the opening of a buffet restaurant at each end of the cantilevered truss a commercial proposition, serving coffee and snacks 52 metres above high water level. By the 1920s, the restaurant had been converted into a café and shop where those souvenir postcards could be bought and date-stamped – or have a colourful sticker attached to commemorate the achievement of having done the climb and the walk.

Although celebrated in a number of contemporary postcards, the 'restaurant in the sky' idea does not seem to have been an innovation which Arnodin repeated. He would, however, surely have approved of the addition of a pedestrian walkway to the Viscaya bridge in the 1990s – more than a century after he and Palacio designed it – and the potential inclusion of café bars high above ground level on the new transporter bridges being proposed at Nantes and Marseille.

For the bridge over the Madelaine branch of the Loire at Nantes, Arnodin explored new design territory, abandoning the tried and tested combination of suspension cables and cable stays for a cable-stayed design with a cantilevered central section. The result was a bridge which, while very similar in span to the Rouen crossing, looked very different.

Lévy, Fils et Cie published this ornately coloured postcard of the Marseille bridge with the gondola midstream, c.1920.

Thanks to local postcard photographers, images survive of the bridge in various stages of construction during 1902 and the early months of 1903. They stand as a testament not only to the golden age of the photographically-illustrated picture postcard – they are very early examples – but also to the fascination with which the local population watched the development of any technological innovation.

The Nantes bridge survived for 55 years before it was dismantled in 1958. All that survives today are the stone piers on both the Quai de la Fosse and Ile-de-Nantes sides of the Madelaine.

A similar design was used for the 165 metres span at Marseille two years later – and that bridge was also fitted with an electric elevator to take visitors to the restaurant and souvenir shop on the beam. Such a novelty was the 'ascenseur electrique' that it featured on its own postcards.

But for the largest of his bridges to be completed, spanning 197 metres across the Usk at Newport, he reverted to a combination of cable stays and suspension cables.

Arnodin is credited with building seven transporters, plus the Bordeaux bridge which was never completed. He was also retained as a consultant on several others – not surprising given the comprehensive patent coverage he and Palacio had acquired.

In his pamphlet *Transporter Bridges*, completed in 1908 but not published until 1912, Henry Grattan Tyrrell said that up to that time Arnodin had: '… erected at least eight of these structures at Bilbao, Bizerta [Bizerte], Rouen, Rochefort, Nantes, Marseilles, Newport and Tangier.'

As the first seven of these have been listed in the chronological sequence in which they were constructed, it might reasonably be assumed that Tyrrell was implying that work on the Tangier bridge had began after 1903, the date at which construction of the Newport bridge started. Sadly, despite extensive research no records of work on such a bridge even being started have been found in either Europe or Morocco.

Until firm evidence of its existence can be uncovered, it must, therefore, remain no more than conjecture that such a bridge was ever built. At the time Tyrrell wrote his booklet, it is highly unlikely he would have visited the location himself.

By the turn of the century Arnodin had suggested several others. His proposals for one across the Seine at Tancarville reached an advance stage, being published in 1897 in *Traversée de la Seine maritime sur transbordeur à la pointe de Tancarville* by Orléans publisher P. Pigelet et Fils.

That would have been his biggest bridge by far had work on it proceeded – the Seine at La Pointe de Tancarville was nearly 500 metres wide.

He actually proposed two transporter bridges across the Seine, the one at Tancarville and six years later a more ambitious cantilevered and cable-stayed bridge at Quillebeuf-sur-Seine, which would have been supported by four towers and able to carry railway trains of up to 225 tons.

Like the proposed Ribble bridge about which Arnodin was consulted in 1900, the proposed Shields bridge across the Tyne did get as far as a formal

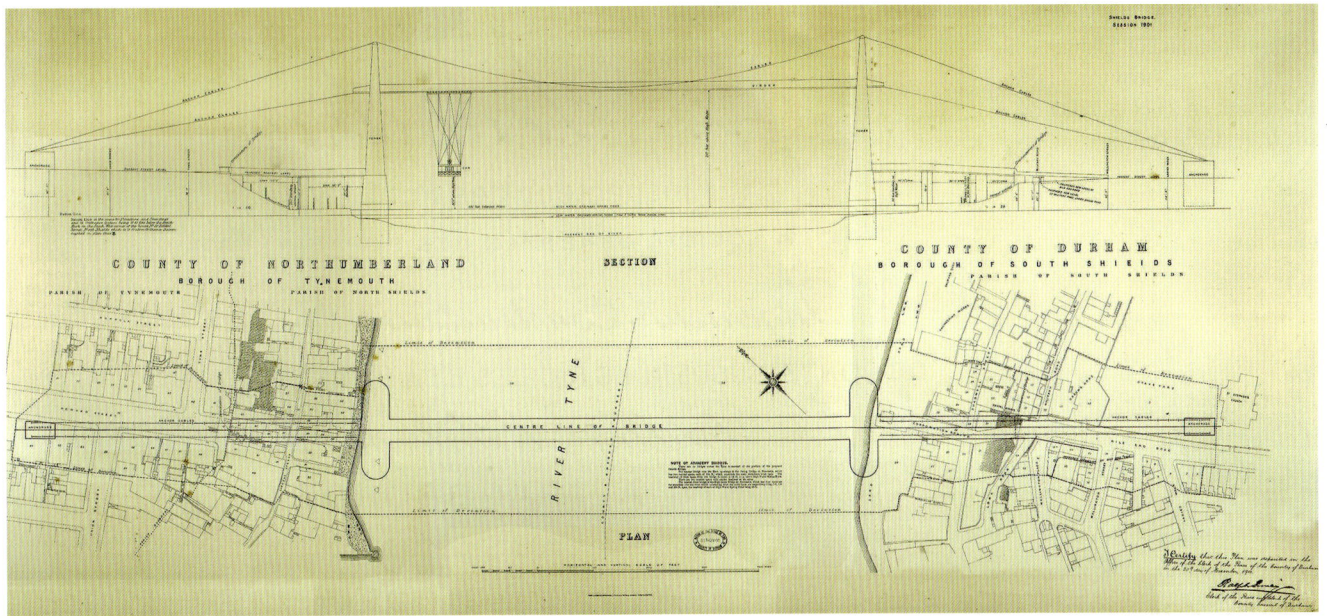

bill being laid before Parliament – an outline proposal and plans having been drawn up by Arnodin and local engineer Charles H. Gadsby. A notice of intent was published in *The London Gazette* on 23 November 1900.

One of the drawings for Arnodin and Gadsby's 1900 Shields Bridge proposal linking Tynemouth and South Shields over the Tyne.

'Notice is hereby given that application is intended to be made to parliament in the ensuing session for an Act to effect the purposed or some of the purposes, following (that is to say):

1. To incorporate a Company (hereinafter referred to as "the Company") and to empower the Company to make and maintain the bridge and works hereinafter described (that is to say):- A bridge across the River Tyne and the banks and foreshore thereof, commencing in the parish and borough of South Shields in the county of Durham at the north-west end of Mile End-road, and terminating in the parishes of North Shields and Tynemouth in the county of Northumberland at the south-east end of Howard-street; such bridge to be constructed and worked in part as a fixed bridge and in part as a transporter bridge with a moveable platform.'

Despite enthusiasm from within Parliament, it was eventually rejected as being too vulnerable in time of war, leading to concerns about the disruption to the factories which, it was anticipated, would develop on both sides of the river if such a bridge was built.

Why the proposed Shields Bridge should have been any more vulnerable than any other bridge across the Tyne was never enlarged upon. Had it ever been constructed, the bridge would have had a span of nearly 255 metres between the towers – 58 metres more than Arnodin's longest completed transporter, the 1906 bridge at Newport.

Its design would have resulted in a bridge fitted with rails on the gondola, enabling the newly electrified Tynemouth & District Electric Traction Company

One of the flyers produced by the Arnodin company in the early years of the twentieth century celebrates his transporter bridges.

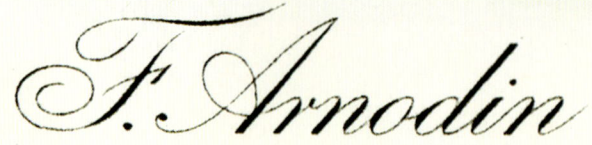

So important was Arnodin factory in Châteauneuf-sur-Loire that postcards of it were published. Here, workers are photographed leaving the works at the end of their shift c.1910.

A. H. - 85 BORDEAUX — Pile en construction du Pont Transbordeur

174 BORDEAUX — Les Quais. — Le Quai transbordeur. — LL

to operate their trams across the river to and join up with the South Shields Tramways Company's network.

Unusually, the gondola would have crossed the river more than 18 metres above high water level. The gondolas on most transporter bridges passed over the water just a metre or two above the normal high water level.

From the souvenir brochure published to mark the opening of the Middlesbrough Transporter Bridge over the Tees in 1911 – an Imbault design although Arnodin was closely involved throughout its construction – an interesting table gives relative costings for nine of the bridges already completed by that time. It listed ten, but assumed that the Pont Transbordeur du Rochefort–Martrou was two separate bridges rather than one.

The accompanying table of construction costs quotes £38,000 for the Puente Viscaya, £36,000 for Rouen, £28,000 for Rochefort and just £24,000 for Bizerte. While no figures are given for either Nantes or Marseille, all these quoted costs, if correct, pale into insignificance when compared with the estimated £98,753 for the Newport bridge, £130,000 for the huge Widnes–Runcorn transporter and £84,000 for the Tees bridge itself.

Above left: One of a series of postcards showing the construction of the towers for the Transbordeur-Médoc. The additional weight of the crane on the incomplete tower was carried by temporary supports.

Above right: One of the completed towers, photographed for postcard publishers Léon & Lévy just before work was halted.

TRANSBORDEUR-MÉDOC

Etablissement d'un Pont à Transbordeur
sur la Garonne, à Bordeaux,
au droit du cours du Médoc.

AVANT-PROJET

DE

CONVENTION DE RÉTROCESSION

Entre les soussignés :
M. Louis Lande, officier de la Légion d'honneur, maire
de la ville de Bordeaux, agissant en cette qualité et
en vertu des délibérations du Conseil municipal en date
des 24 mars 1902 et
D'une part;

Et M. Ferdinand Arnodin, chevalier de la Légion
d'honneur, ingénieur-constructeur, demeurant à Châteauneuf-sur-Loire (Loiret),
Agissant en son nom personnel et sous l'inspiration
du Comité de patronage du Pont à Transbordeur de
Bordeaux, constitué sous la présidence de
M. Charles Cazalet, officier de la Légion d'honneur,
ancien adjoint au maire de Bordeaux, et composé de :
M. Adrien Bayssellance, officier de la Légion d'honneur, ancien maire de Bordeaux, ingénieur en chef de
la Marine en retraite, vice-président;

TRANSBORDEUR-RICHELIEU

Établissement d'un Pont à Transbordeur
sur la Garonne, à Bordeaux,
au droit de la place Richelieu.

AVANT-PROJET

DE

CONVENTION DE RÉTROCESSION

Entre les soussignés :
M. Louis Lande, officier de la Légion d'honneur, maire
de la ville de Bordeaux, agissant en cette qualité et
en vertu des délibérations du Conseil municipal en date
des 24 mars 1902 et
D'une part;

Et M. Ferdinand Arnodin, chevalier de la Légion
d'honneur, ingénieur-constructeur, demeurant à Châteauneuf-sur-Loire (Loiret),
Agissant en son nom personnel et sous l'inspiration
du Comité de patronage du Pont à Transbordeur de
Bordeaux, constitué sous la présidence de
M. Charles Cazalet, officier de la Légion d'honneur,
ancien adjoint au maire de Bordeaux, et composé de :
M. Adrien Bayssellance, officier de la Légion d'honneur, ancien maire de Bordeaux, ingénieur en chef de
la Marine en retraite, vice-président;

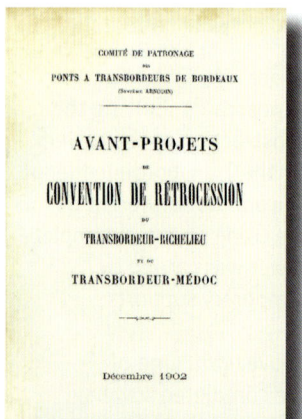

COMITÉ DE PATRONAGE
des
PONTS À TRANSBORDEURS DE BORDEAUX
(Section ARNODIN)

AVANT-PROJETS
de
CONVENTION DE RÉTROCESSION
du
TRANSBORDEUR-RICHELIEU
et du
TRANSBORDEUR-MÉDOC

Décembre 1902

His grandest scheme, however, would turn out to be his last. It was also the only one to be abandoned when partially built.

Much has been written over the past century about Arnodin's proposal for a bridge across the Garonne at Bordeaux which would have had a span of 430.3 metres, almost exactly the same as J.W. Morse's 1869 speculative and impractical proposal for his aerial bridge over New York's East River.

But the true story of Arnodin's plans to bridge the Garonne is much more complex than simply a revolutionary bridge design which was never realised.

Henry Gratton Tyrrell in his 1908 book had obviously misunderstood the proposal, and published only part of the story. Arnodin was, in fact, proposing two revolutionary bridge designs – one of which would have abandoned all his earlier experience and practice and resulted in a transporter bridge like none built before.

An important document published in 1902 reveals that he proposed not one but two 'transbordeurs' across the river.

They were to be known as the 'Transbordeur-Médoc' and the 'Transbordeur-Richelieu', and they were to be very different in both size and design.

The Transbordeur-Médoc was to cross the river between the Quai des Queyries on the east side of the river and Cours du Médoc on the west, while the second bridge, the Transbordeur-Richelieu, would span the similar width between the Quai de Bacalan near the Place Richelieu – also on the west bank – and the Quai des Queyries, about half way between the Transbordeur-Médoc and the early nineteenth century Pont de Pierre which still stands there today.

Discussions about bridging the Garonne across the navigable channel had originally been initiated in 1893 – around the time Palacio and Arnodin completed the Viscaya bridge – but little progress had been made.

The revival of the idea in 1901 seemed to herald the rebirth of the project, but on a grander scale than had been considered nine years earlier.

Increases in vehicle traffic had already suggested that one crossing would probably be insufficient.

The Pont de Pierre had been built across the river upstream of most of the docks, but the fact that there was no way across the river except by ferry during its long meander north-west towards the sea was a growing problem, and by December 1902 Arnodin, recognising that nobody had ever bridged such a wide river with a single span before, understood it would require a very different design to those employed on his earlier bridges.

Above: The surviving piers of the Transbordeur-Medoc today. The bridge anchor building stands about 200 metres behind the piers.

Opposite: The documents outlining Arnodin's plans for the two Bordeaux bridges were largely concerned with the management structure of the projects, the terms under which staff would be employed, and the legal permissions which would need to be granted. The documents also included details of the terms and conditions under which the bridges would be built, limiting activity on the sites to ten hours per day during construction.

Above left: The completed towers of the Transbordeur-Médoc seen across the city's docks.

Above right: President Fallières arrives in Bordeaux to lay the foundation stone.

The company that would build and hold the concession to operate the bridges was to be known as the 'Societé anonyme des Ponts á Transbordeur de Bordeaux (Système Arnodin)' and would have to raise capital of at least four million francs.

The solutions Arnodin proposed were highly original, showing he had lost none of his ability to embrace highly innovative design solutions to the challenges which faced him.

The 1902 pamphlet offered some tantalising information about his working brief for the proposed Transbordeur-Richelieu, and its intended scale and capacity, but as it was never built, we only have Tyrrell's illustration in *Transporter Bridges* to give us a glimpse of what it might have looked like – and both his descriptive text and the dimensions he included clearly actually bore no relationship to the bridge on which work would eventually be started.

While the two bridges would have had very different carrying capacities – and be built to radically different designs – both would have required spans in excess of 400 metres.

'Flower girls' took part in the laying of the foundation stone for the Transbordeur-Médoc during the major celebrations known as the 'Fêtes Presidential' in September 1910.

59. BORDEAUX — Fêtes Présidentielles - Les quatre jeunes Filles qui offriront des fleurs, le marteau et la truelle pour la Cérémonie de la pose de la première pierre du Pont transbordeur " Médoc " - M. D.
Marcel Delboy. Phototypie - Bordeaux

83. BORDEAUX
Le Transbordeur en construction
BR - 402

'La Ville de Bordeaux s'engage à demander immédiatement à l'Etat la concession d'un pont à transbordeur à plusieurs nacelles, à construire sur la Garonne à Bordeaux, au droit de la place Richelieu, et cela aux conditions d'un cahier des charges dont le projet est annexé au present traité.'

One of the Bordeaux towers standing on the quayside at the Cours du Médoc.

53. - BORDEAUX. - Les Pylônes du Pont transbordeur

Another of the many postcards produced in the interval between the towers being completed and the project being halted by the outbreak of war.

Translated:

'The City of Bordeaux agrees to apply immediately to the State for the concession for a multi-nacelle ferry bridge to be built on the Garonne in Bordeaux, at the Place Richelieu, subject to the conditions in the attached book of draft proposals.'

So what was initially planned would have gone a long way towards addressing the ever-increasing traffic volume, cutting waiting times by having two nacelles, or gondolas, able to pass mid-stream, thus doubling the bridge's capacity.

As has already been mentioned in an earlier chapter, a twin-gondola proposal would be revisited by the Dublin Docks & Harbour Board in 1929 for a transporter across the Liffey. Had this bridge ever been built, it would have been a remarkable structure as Arnodin was proposing such a radical and untested design. But it never got beyond the drawing board. Tyrrell was not to know that, and in his 1908 treatise described the only Bordeaux transporter bridge which he knew about – but work on such a bridge, of course, had not yet even reached the detailed planning stage at the time he was writing in 1908.

'The most daring project for a transporter bridge ever undertaken was that which appeared in 1903 for crossing the Gironde River at Bordeaux, with a single arch of 1,412 feet the span being about the same as that designed by Morse for crossing the East River in New York. The proposed Bordeaux bridge consisted of a pair of metal arches, in vertical planes and about 80 feet apart, from which the runway deck was suspended, leaving a clearance of 150 feet beneath it. The total rise of the arch was 328 feet … The clear distance between docks was to be 1,312 feet, making it longer than any arch yet built. The runway deck had provision for a footwalk but was without stiffening trusses, and it supported a double line of track so that cars might start from each side of the river at the same time.'

With hindsight – and with reference to the marked differences between the two proposals described in the 1902 pamphlet – it is clear that he was writing about the proposed Transbordeur-Richelieu – on which construction work was

The design for the Transbordeur-Médoc suggested using what Tyrrell described as 'lune-shaped ribs' to create a more rigid structure from which to hang the stiffening beam. Given the width of the proposed span, this design would have flexed under loading to a significantly lesser degree than a conventional cable suspension structure. The stone piers and cable anchors on the east shore still survive.

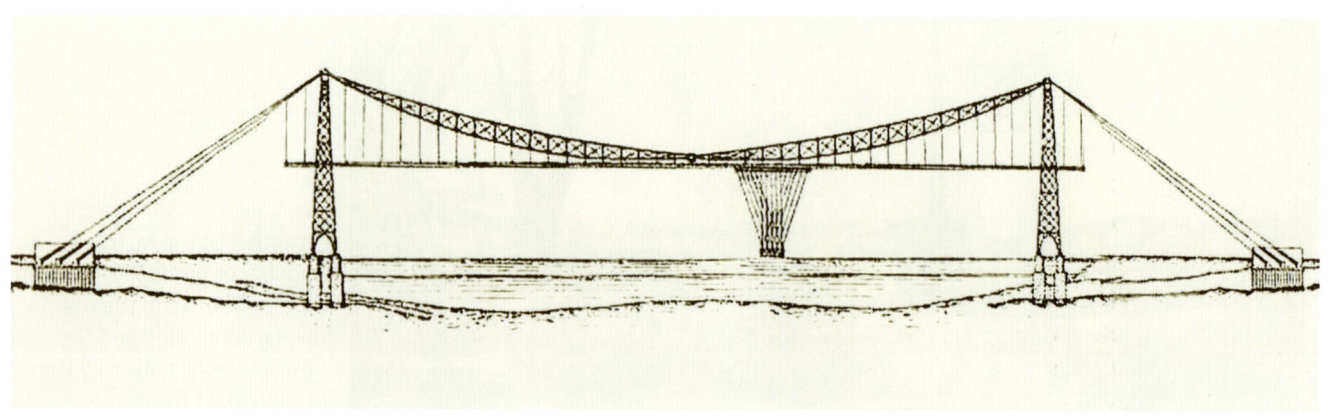

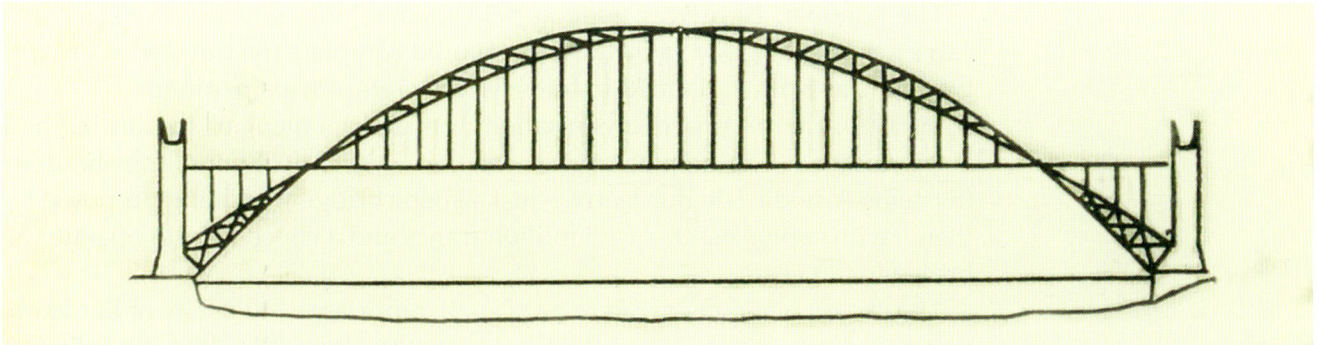

never started – and not the Transbordeur-Médoc as has been widely assumed over the past century.

Whether or not this would have created a stable structure across such a wide span we will never know, for there is no evidence that detailed design and planning for it was ever started.

For the Transbordeur-Médoc Arnodin reverted to a design of two tall pylons – but using inverted hinged metal arches in place of the main cables in the belief that such a design would minimise the natural tendency of such a wide beam to deform under load.

The 1902 proposal contained several different options – each requiring the resolution of different engineering challenges, suggesting that a final design had not yet been fully articulated.

So, for the second bridge:

'La Ville s'engage à demander immédiatement à l'Etat la concession d'un pont à transbordeur à une ou plusieurs nacelles à construire sur la Garonne à Bordeaux, en face le couers du Médoc, et cela aux conditions stipulées dans le cahier des charges qui sera définitivement adopté pour le premier transbordeur de la place Richelieu.'

Translated:

'The City undertakes to apply immediately to the State for the concession for a ferry bridge of one or more nacelles to be built on the Garonne in Bordeaux, opposite the Couers du Médoc, subject to the conditions stipulated in the specifications which will definitively be adopted for the first ferry bridge at Place Richelieu.'

Clearly a decision had yet to be made as to whether the second bridge would be designed to operate with 'one or more' gondolas. Further delays meant that actual construction of the bridge did not start until 1910 – 17 years after the project had first been discussed.

By the time work started, it is clear from the profile of the towers – which had been completed when the First World War brought the project to a halt – that the bridge was to be of a single gondola design and very different to that described by Tyrrell.

Tyrrell's sketch of Arnodin's proposed Transbordeur-Richelieu. According to Arnodin's specification, this bridge would have involved a total re-imagining of the transporter bridge. In 1906, an arched transporter bridge was one of the options considered by Middlesbrough Corporation for their Tees bridge.

The gondola, 10 metres wide and 13 metres long, had been designed to carry a maximum load of 50 tons, and would complete the traverse in around two minutes with a planned frequency of six return trips per hour.

By 1910, the proposed organisation and management of the bridge had been modified. The new company was the 'Societé du Pont á Transbordeur (Systéme Arnodin) de Bordeaux' – just a single bridge was being proposed by then – with capital assets of 1.5 million francs and a concession to operate the bridge for 80 years.

The project was considered to be of such importance to the city of Bordeaux that no lesser personage than the President of the Republic, Armand Fallières, was invited to lay the foundation stone as part of major celebrations known as the 'Fêtes Presidential' in September of that year.

Work on the towers progressed through 1911 and 1912, their final appearance being very similar to tried and tested Arnodin designs, albeit on a much larger scale, but progress was halted during the First World War, and never restarted after hostilities ceased. The towers subsequently stood, abandoned and unconnected, until 1942 – becoming well-known Bordeaux landmarks in their own rights – after which they were dismantled and their steelwork recycled as part of France's war effort.

Only two of their stone plinths survive today by the water's edge along the Quai des Queyries – across from the river cruisers' quay. The stone and concrete cable anchor block stands around 200 metres back from the tower bases. It would not be until 2013 that this part of the Garonne was finally bridged – by the Pont Jacques-Chaban Delmas lift bridge whose 77 metre high towers dominate the river.

Monsieur Arnodin was spared the sight of seeing any of his bridges being demolished, having died in 1924. Alberto Palacio outlived him by 15 years, dying in 1939.

All of Arnodin's bridges outlived him, providing a slow, but stately, means of crossing rivers if time was of no concern, but as queues of vehicles grew longer, high level road bridges were built to replace many of them.

The bridge across the harbour mouth at Brest, damaged in 1944, was demolished in 1947; the Rouen bridge was a casualty of war in 1940, as was Marseille in 1944. The Nantes and Rochefort bridges continued in use long after hostilities ended, Nantes being demolished in 1958.

The Pont Transbordeur de Rochefort–Martrou survived as the only local crossing of the Charente until the Pont levant de Martrou opened in 1966, after which the transporter was decommissioned. The story of its 'rebirth' is told in the final chapter of this book.

The age of the transporter bridge had passed long before either Arnodin or Palacio did, but the three surviving bridges which they created together and singly – at Portugalete, Rochefort and Newport – are enduring reminders both of their vision and their engineering prowess.

The Portugalete bridge has already been enrolled as a World Heritage Site by UNESCO. All the world's other surviving transporter bridges deserve similar recognition.

Above: Bordeaux's 2013 Pont Jacques Chaban–Delmas lift bridge – Europe's largest at 575 metres – crosses the river 400 metres downstream from the never-completed Transbordeur-Medoc.

Left: Two of Arnodin's bridges – the Pont Transbordeur de Rochefort and the Pont à Transbordeur de Marseille – are remembered in collectors' medallions struck by 'Monnaie de Paris' in 2010, 2012 and 2016, the most recent recognising that the Rochefort bridge is the last surviving example in France.

BUILDING THE WIDNES-RUNCORN BRIDGE

From an historical perspective, it is unfortunate that the most magnificent of all the transporter bridges ever built – and the longest – no longer survives, having been closed and demolished in 1961. Although officially known as the Widnes–Runcorn Transporter Bridge at the time of its construction

Opposite and left: Public interest in the building of the Widnes and Runcorn Transporter bridge was considerable. Runcorn District Council published tinted postcards showing the progress of the work. This view of the Runcorn tower is a detail from a postcard published at the end of April 1904. By then the tower was complete, the main cables in place, and the construction of the stiffening beam under way. As the full postcard shows, shipping continued to use the Manchester Ship Canal as construction progressed.

WIDNES & RUNCORN TRANSPORTER BRIDGE. Opened by Sir John T. Brunner, M.P., May 29th, 1905. W. H. MACK'S SERIES.

Taken from the Widnes shore, this postcard by W.H. Mack of Runcorn commemorated the opening of the bridge on 29 May 1905.

Sinking the cast-iron cylinders which supported the bridge, 23 June 1902. The L&NWR's 1868 railway bridge can be seen behind.

and opening, for several years thereafter, it was more often referred to as the Runcorn Transporter Bridge in Runcorn and the Widnes Transporter Bridge in Widnes. From 1911, when ownership was transferred by Act of Parliament to Widnes Council, it was officially the Widnes Transporter Bridge.

A worker entering the air-lock on 1 September 1902 within which navvies excavated the river bed while sinking the huge cylinders – a dangerous and unpleasant job. Sinking the cylinders was sub-contracted to Holme & King of Liverpool.

RUNCORN BRIDGE (OVER THE MERSEY.)

The 1868 Runcorn Railway Bridge, from a Raphael Tuck postcard published in 1903 for the London & North Western Railway. Before the transporter bridge was opened, the pedestrian walkway on the railway bridge was the only means of crossing over the River Mersey and the Manchester Ship Canal at this point, apart from the L&NWR's small ferry.

One can usually, but not always, tell when and where a postcard was published by the name given to the bridge – even the magazine articles describing progress in its construction occasionally failed to agree on its name.

With a span of nearly 304 metres and towers which rose 60 metres above high water level, at the time of its opening in 1905 it was almost twice as long as any transporter bridge yet built anywhere in the world. It was designed by Liverpool-based engineer John T. Wood and Londoner John J. Webster for the Widnes & Runcorn Bridge Company.

Work in progress on the bridge piers on the Widnes side of the river, Christmas Eve 1902.

THE CONSTRUCTION OF THE WIDNES AND RUNCORN TRANSPORTER BRIDGE

(For description see page 444)

Fig. 1—MARCH 5th, 1903

Fig. 2—FEBRUARY 24th, 1904

Fig. 3—SEPTEMBER 23rd, 1904

Fig. 4—MARCH 31st, 1905

MAY 5, 1905

THE ENGINEER

445

A page from the 5 May 1905 issue of *The Engineer* showing progress photographs of the building of the bridge, including, *top left*, the earliest of the series from March 1903 as the caisson for the Widnes tower was being constructed.

The proposed bridge's scale made it a massive feat of engineering, the iron and steelwork for which was entrusted to the Arrol Bridge and Roof Company of the Germiston Works in Glasgow. Founded in 1883 by two brothers J. Cameron Arrol and J. Arthur Arrol, this company should not be confused with that other Glasgow engineering enterprise, the more famous Sir William Arrol & Company of Dalmarnock. In a Glasgow trade directory of 1888 – listing Arrol Brothers as the entry before their more famous namesake – the compiler had noted that

'The steel and iron trade of the Glasgow district has now attained a position of great importance and value among the industries of the country, while the many sections of the trade are remarkable for their variety and application. Large concerns have put down enormous plant for special purposes, and in this respect Messrs. Arrol Brothers, of the Germiston Works, in the north eastern suburb of Springbum, have distinguished themselves for their devotion to the particular work of steel and iron bridge and roof building. In an ordinary way this would appear to be an industry sufficiently large and distinct enough, in all conscience, but in the iron and steel trades every section or department is of gigantic and mammoth dimensions, and other branches of the general engineering trade also find a place in the Germiston

The Runcorn cable anchorages with temporary adjustment gear still in place, 11 May 1904.

448   THE ENGINEER   MAY 5, 1905

THE WIDNES AND RUNCORN TRANSPORTER BRIDGE
(For description see page 444)

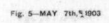

Fig. 5.—MAY 7th, 1903

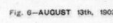

Fig. 6.—AUGUST 13th, 1903

Fig. 7.—AUGUST 25th, 1904

Fig. 8.—FEBRUARY 10th, 1905

Also from the 5 May 1905 issue of *The Engineer,* key stages in the construction of the Widnes tower in the 15 months between between May 1903 and August 1904, and one view from the Runcorn side as the project neared completion in February 1905.

Works … Amongst other contracts which this firm has executed since they started are the 70 ft. diameter caissons for the Forth Bridge, in which there are upwards of 2,000 tons of iron and steel.'

Not surprisingly, the project generated a lot of interest both in specialist engineering journals and in populist media during its construction. Postcards were regularly published by several local photographers as the work progressed, amongst them W.H. Hall of Princes Street in Widnes and W.H. Mack in Runcorn.

The journal *Engineering* published its first account of the bridge – with a brief historical context – a few weeks before it was opened in 1905, illustrated with Mack's detailed views of the bridge's construction and its equipment.

'The idea of replacing an ordinary bridge by an aerial ferry of this character is not, of course, new, having been proposed some thirty years ago by Mr. Charles Smith, of Hartlepool. A small one was erected at the Devil's Dyke, near Brighton, a few years since, as an attraction for trippers, and large

Opposite: From the 16 June 1905 edition of *Engineering*, the interior of the power house with its pair of 75bhp Crossley Gas Engines. The switchboard can be seen back left. The engines also charged an array of 245 chloride batteries in the room above which provided back-up power. The power house was the tall tapered building which stood within the base of the Widnes east tower.

Left: This illustration from the 16 June 1905 issue of *Engineering* shows one of the two Mather & Platt 35bhp electric motors which drove the travelling frame from which the gondola was suspended.

The first official crossing, the gondola laden with dignitaries.

The bridge's two main engineers John T. Webster, of Westminster, London, *top*, and John T. Wood of Liverpool.

bridges on the same system have also been built over the Seine, at Rouen; over the Nervion, in Spain; and at Bizerta, in Algiers. The object in each case has been to reduce the high capital expenditure inseperable from any other system of freeing a navigable waterway. In the first instance, the total expenditure, including Parliamentary expenses, has been 130,000l.; a high level bridge would have cost three times as much, and a subaqueous tunnel still more.

The bridge has a clear span of 1000 ft. As shown in our illustrations prepared from photographs by Mr. W.H. Mack, Runcorn, the bridge is of the stiffened suspension type, the cables being 12 in. in diameter, built up of steel wire, having a tensile strength of 95 tons per square inch. These are anchored into solid rock at each abutment. The towers are 160 ft. high. The stiffening girders are 18 ft. deep, and are spaced 33 ft. apart. They are braced together horizontally. Their lowest points are 82 ft. above high-water level. Rails carried by these girders support the trolley, from which the car is suspended. The trolley is mounted on thirty-two wheels, and is 77 ft. long. It will be propelled by electro-motors, and will take 2¼ minutes to make the trip from bank to bank. The car suspended from this trolley will accommodate 300 foot-passengers, in addition to four two-horse farmers' wagons. The small generating station is provided on one bank, to supply the current needed for operating the trolley. Until the construction of this bridge, there was none available for vehicular traffic between Warrington and the sea, though at Runcorn a footpath on the London and North-Western Railway Bridge was open for foot passengers.'

The Devil's Dyke 'bridge' referred to was, in fact, an aerial cableway – what we might think of today as a cable-car – and therefore not considered a transporter bridge within the definition accepted for this project, but the importance of the Widnes and Runcorn bridge in aiding the fortunes and development of both towns cannot be over-estimated.

As that quotation mentions, prior to the bridge's opening, pedestrian access across the river had involved a considerable walk across the Britannia Bridge, the railway viaduct which had opened across the Runcorn Gap in 1868. The terms and conditions under which the London & North Western Railway had been granted powers to build the bridge had stipulated that a pedestrian footway must be part of the design.

Long after the transporter bridge opened, that walkway – now closed off – provided photographers with a great vantage point to photograph the gondola as it traversed the river.

The transporter bridge was popular from the day it opened, stimulating cross-river trade between the two towns – so popular that quite regularly there were waiting queues of traffic. At either end of the bridge, small buildings of local red sandstone were erected to house ticket offices and waiting rooms for pedestrians, while, on the Widnes side of the river, the company's offices – which still stand, although now adapted for other uses – occupied a corner site overlooking the bridge on Mersey Road.

THE RUNCORN AND TRANSPORTER BRIDGES. WIDNES. PHOTO BY F.H.HOWELLS.

Local photographer F. H. Howells of 128 Widnes Road, Widnes, produced this postcard of the transporter bridge and the walkway on the railway bridge in the late 1920s or early 1930s. He had operated a photographic studio in the town since before the First World War.

This illustration from the 7 April 1905 issue of the journal *Engineering* shows the Widnes towers linked to the shore by a cast-iron bridge, with the power house set within the base of the east tower.

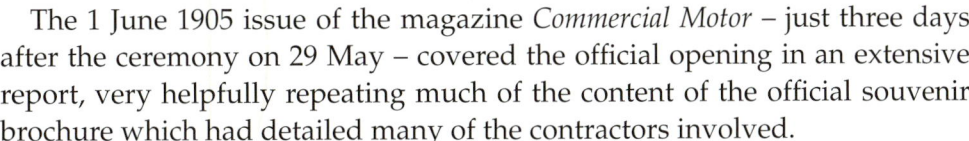

Opposite and above: Members of the public have their first experience of crossing the river by 'electric car'. The postcard, by W. Hall of Widnes, was available as a photographic print, or a tinted postcard.

Above left: A multi-view postcard from 1912.

Left and below: The former Transporter Bridge Offices building still stands at the foot of Mersey Road, Widnes.

The 1 June 1905 issue of the magazine *Commercial Motor* – just three days after the ceremony on 29 May – covered the official opening in an extensive report, very helpfully repeating much of the content of the official souvenir brochure which had detailed many of the contractors involved.

As a pre-amble to the report, the writer recalled that the great Scottish engineer Thomas Telford had designed a bridge to cross the river at exactly the same point more than a century earlier, but despite successfully demonstrating the viability of a chain suspension bridge with the large span necessary to cross the river, it had never got off the ground.

Above: The gondola as illustrated in the account of the bridge's construction in the journal *Engineering*.

Right: Photographed at low tide, the approach ramp can be seen to be of a similar construction to a seaside pier. The design of the superstructure upon which the power-house was built would cause significant operational problems which were not resolved until the direct DC-drive system was replaced in 1913, and use of the Crossley gas engines discontinued.

'The engineers for the work are Mr. John J. Webster, M.Inst.C.E., of Westminster, and Mr. John T. Wood, M.Inst.C.E., of Liverpool; the resident engineer being Mr. L.H. Chase, M.Inst.C.E. The contract for the masonry in the approaches and anchorages was let to Messrs. W. Thornton and

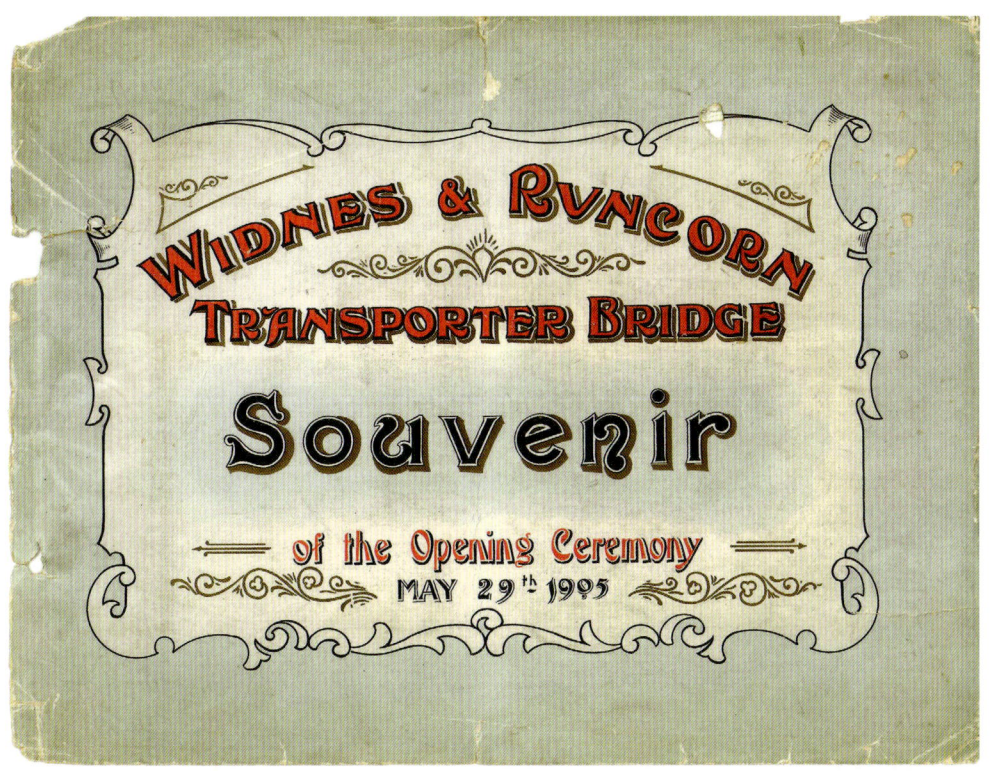

The front cover of the illustrated Souvenir Brochure – which cost the princely sum of one shilling – produced to mark the bridge's opening.

Sons, Liverpool; the contract for the steel super structure being let to Arrol's Bridge and Roof Company, Glasgow, who sub-let the sinking of the cylinders, green heart fenders and scaffolding to Messrs. Holme and King, Liverpool. The construction of the towers, approach girders and cylinder foundation was sub-let to the Widnes Foundry Company. The construction of the steel cables was let to the St. Helens Cable Company.

These transfer-printed souvenirs amongst the many commemorative items produced to mark the opening of the bridge in 1905. Neither bears a maker's mark. The little cup measures just 70mm in diameter, while the mug stands 85mm high.

Opposite above: Arriving at Runcorn, by W.H. Hall.

Opposite below: The transporter bridge in the 1930s The building on the bridge was the winch-house built in 1913 when the DC motors were replaced by a winch drive system. The cables which can be seen going up to the stiffening beam enabled the car to be driven either from the winch house or from the gondola itself.

Below: One of the cable saddles which carried the cables over the tops of the towers. The saddles allowed the cables to move due to expansion and contraction or changes in the stresses on the main beams as the gondola crossed the span.

The whole of the electric installation and equipment, including the lighting of the structure, was carried out by Messrs. Mather and Platt, of Salford Ironworks, Manchester.'

The report ended with some information for users of the bridge.

'It was announced during the course of the proceedings that the L. and N.W. Railway would now discontinue the boat ferry which they had been obliged to run under Act of Parliament. This ferry had been in existence for over 700 years. The toll for each foot passenger would be 1d., and 6d. for motorcars and other carriages, providing the weight did not exceed one ton. For heavier vehicles, with a maximum of 13 tons, the charge would be 6d, for the first ton, and 9d. a ton for each additional ton. These rates certainly are unduly heavy in respect of a 5-ton wagon.'

While both *Engineering* and *Commercial Motor* accompanied their articles with illustrations of the bridge either nearing completion or complete – *The Engineer*, in its issue for 5 May 1905, publishing a remarkable series of photographs showing the progress of the construction, the earliest from March 1903 when work on the caissons which would support the towers was well underway.

The article explored the challenges facing the engineers, both in establishing strong enough foundations for the towers, and achieving the maximum rigidity in the stiffening girders – for they were, in engineering terms, two

separate girders joined by cross braces. Of the towers and their foundations, the report said:

'Each tower consists of four members or legs of open square section, 4ft. 10in. wide at the base, tapering to 2ft. 3in. wide at the top, all well braced together with strong horizontal and diagonal bracing. The four legs of each tower are spaced 30ft. apart at the base and curve inwards as they ascend till they reach the top landing where they are 6ft. 9in. apart. Each pair of towers are 70ft. apart, and are braced together by arched framing at the top, and by strong horizontal and diagonal bracing lower down. All four members of each tower are bolted at their base to cast iron cylinders, 9ft. in diameter, which in their turn are securely bolted to solid rock foundations. On the Widnes side the rock is near the surface, but on the Runcorn side it is some 35ft. below the Ship Canal level, and here the cylinders were sunk by means of compressed air and filled with strong cement concrete. The bases of the towers are protected by massive timber fenders secured to piles.'

WIDNES TRANSPORTER BRIDGE. (6)

209784.J.V.

A view along the stiffening beam showing the lattice girder construction.

Before describing how the 12-inch thick steel cables were made up – from 19 separate 'ropes' each of which was itself made up from 127 steel wires, the whole being wrapped in bitumen-soaked sail cloth for weatherproofing – the writer focused on some of the systems which the designers hoped would ensure the stability of the structure under normal atmospheric conditions.

'Upon the top of the towers are saddles mounted on steel rollers for carrying the two great steel cables from which the stiffening girders are suspended. These rollers are adapted for taking up the variations in the length of the cables due to the load and temperature.'

Whereas a conventional suspension bridge can be designed in such a way that the total weight of the suspended carriageway and its traffic can be spread evenly along its span, transporter bridges pose a different and distinctive problem. Although the traffic load is less, its weight is concentrated at any one time on a very small section of the suspended beam. Thus, in normal use, there will inevitably be a flexing of the suspension cables and girders as the gondola moves across the river. Designing a structure in which those deviations were minimised and manageable, and which offered a smooth and reliable crossing over a span of hitherto untried length, posed significant challenges.

With a bridge with such a considerable span as the Widnes–Runcorn transporter, wide towers were a key component of the design, with the stiffening beam sitting within them. This ensured that there was adequate cradling of the beam to maintain a relatively stable structure.

To move the gondola across the river, a cable-winch system was initially discounted as unreliable due to the length of the span, the designers opting instead for a completely untried system for powering the travelling frame.

However, the chosen direct drive system brought with it its own challenges. With DC motors directly driving the overhead travelling frame wheels, and power being picked up from a live third rail, engineering a design that maintained good electrical contact was vital. The souvenir brochure explained how it all worked:

'It [the 'trolley' or 'travelling frame'] is propelled by two electric motors of about 35 brake horse-power each, a large excess of propelling power being provided for economy of working and principally to be ready for any emergency of strong head wind. The motors are fixed to a kind of bogie arrangement in the trolley, so that in the case of large curvature of the bottom boom of the stiffening girders, due to either temperature or load, the driving wheels will be certain to be hard on the rails.'

HMS *Bristol* passing under the bridge, probably in 1911 – making her way along the Manchester Ship Canal, *en route* towards Manchester Docks. As this same picture appears in a photo-composite – produced around 1915 – which includes four images which have been identified as the work of W. Hall of Widnes, this fine study is likely by him as well. HMS *Bristol* had been completed at John Brown's yard on the Clyde and handed over to the Royal Navy in December 1910.

H.M.S. BRISTOL, PASSING UNDER THE WIDNES TRANSPORTER SPAN.

Widnes and Runcorn Transporter Bridge

Benbow Series.

A tinted postcard published around the time of the bridge's opening in 1905.

The electricity was generated by two 75bhp Crossley gas engines housed in a power generating station built within the legs of the Widnes east tower. The engines were driven by town gas which was piped in under the approach piers from Widnes Municipal Gasworks.

A significant problem was encountered as a result of the positioning of the gas engines. The power-house under the Widnes east tower was not an especially robust structure, which in turn meant that the engines were not mounted on a firm enough foundation – their position was, in many respects, akin to being at the end of a seaside pier. As a result, they caused the structure to vibrate – and that could even be felt by passengers waiting to embark at Runcorn.

In times of light use, any excess electricity generated was used to charge 245 chloride cells in the room above the engines, and this stored power was used to help out at times of great demand, and also to produce the electricity used to light the bridge and its gondola.

The 220 volt DC current, in turn, powered a pair of 35bhp Mather & Platt motors directly mounted on the bogies of the overhead carriage, and that propulsion system proved to be woefully unreliable. With a fully laden gondola, the flexing of the stiffening beam caused intermittent loss of contact between the electrical pickups on the carriage and the overhead rail as the traveller made their way across the river and the ship canal.

By 1913 a decision was made to abandon the integral DC motors driving the bogies on the travelling frame and replace them with a conventional electrically-driven steel cable and winch system. A winch-house was built over the roadway at the Widnes end and this continued to provide drive until the bridge was closed in 1961.

BOURNVILLE COCOA
MADE BY CADBURY.

TRANSPORT

A SERIES OF 25

14

Transporter Bridge, Runcorn.

This type of bridge solves the problem of transporting passengers and goods across a river without interfering with shipping, and without incurring the expense of a high-level fixed bridge. The horizontal girders are supported by steel towers, and carry on their under side rails along which a trolley can travel. From this trolley there hangs a car, supported by vertical cables, and driven by electric motors fixed to the trolley. One of the largest Transporter Bridges crosses the Mersey and the Manchester Ship Canal between Widnes and Runcorn, and has a span of 1,000 feet. It is 82 feet above high water.

SEE THE NAME "Cadbury" ON EVERY PIECE OF CHOCOLATE

A Cadbury Collectors' Card titled 'Transporter Bridge, Runcorn', describing the bridge as it was pre-1913, and a British Transport Commission ticket from the 1950s. The bridge is identified, here too, as the 'Runcorn Bridge' rather than the 'Widnes Bridge' or 'Widnes–Runcorn Bridge' as it had been under Widnes Council control.

Right: Looking the worse for wear, the gondola part way across the river, photographed from the walkway on the railway bridge in 1959. At that time it cost a penny to use the railway walkway – plus an additional ha'penny if you were pushing a bicycle – but it was free to railway employees.

Below: The 1933 Clyde-built SS *Mancunium*, Manchester Corporation's 'sludge boat', passes under the Runcorn/ Widnes bridges in 1959 on its way to dump its cargo at sea. Work on the road bridge which would replace the transporter was already underway and can be seen beyond the left-hand arch of the railway bridge.

Despite its teething problems, considerable interest had been shown in the bridge on the other side of the Atlantic, amidst long-held belief that there were obvious locations where such bridges might be built in America,

That such a huge expanse of water had actually been bridged seems to have renewed interest in the idea – after all an account of J.W. Morse's ideas for a transporter bridge across New York's East River had been published in *Scientific American* 36 years earlier, and while that was ultimately proven to have been a flawed and totally unworkable solution to the problem, the idea had certainly not gone away.

The American magazine *Popular Mechanics* published an account of the completed bridge in their June 1905 issue, containing some fascinating technical details and illustrated with another of Mack's photographs.

'An aerial ferry or transporter bridge, 1,000 feet long in the clear, and spanning the Mersey has been opened to service connecting Widnes and Runcorn. The bridge has the longest span of any bridge in the United Kingdom designed for road traffic. The overhead truss-work is hung from two cables each containing 2,413 wires and weighing 243 tons. The truss is 18 ft. high by 35 ft. wide and allows a clearance of 82 ft. above high water, and is fixed to vertical rockers at each end to provide for expansion and contraction. The towers are of steel, 190 ft. high and rest on eight cast iron cylinders 9 ft. in diameter which are anchored to the solid rock.'

Popular Mechanics estimated the total cost of the Widnes–Runcorn Bridge at $650,000 – given the exchange rates of the day, around 30 per cent more than the £130,000 quoted in both *The Engineer* and *Engineering* – but the latter was just as guilty of inaccuracy having stated that the towers were 160 feet high (51 metres) high, 30 feet shorter than they actually were.

Back on the Mersey, however, to add to the mounting operational problems, the Widnes & Runcorn Bridge Company was quickly in financial difficulties – despite the bridge's popularity and heavy usage – their cash reserves and business plan both proving woefully inadequate. They eventually admitted defeat just under 11 years after the company had been established in 1900 under the chairmanship of Sir John Brunner, and just six years after the bridge had been opened.

From the outset, the management appears to have been unable to generate sufficient income to cover the bridge's operational and maintenance costs and, close to insolvency by 1911, the company was put up for sale and rescued by Widnes Borough Council.

Once under local council ownership, the bridge's usage continued to rise, reportedly peaking at around a million foot passenger crossings a year, as well as being used by an annual quarter-of-a-million vehicles.

Interest in the bridge continued, with new postcards still being published well into the 1930s.

Right: A ship on the Manchester Ship Canal – separated from the Mersey by a concrete wall – passing beneath the bridge in the late 1950s.

Opposite above and below: The early stages of the demolition of the bridge in late 1961, carried out against the backdrop of the bridge's replacement.

Below: Now known as the Silver Jubilee Bridge, the replacement for the transporter bridge opened in 1961, and at the time was the third longest steel arch bridge in the world. The opening of the new bridge was immediately followed by the transporter bridge's closure. This view was taken from the surviving Widnes approach ramp for the transporter bridge.

Amongst the more unusual items to carry an image of the bridge was a 'Collectors' Card' published by Cadburys, and given away inside packets of Bournville Cocoa. The company started producing these cards around 1909, and while the date of the bridge card is usually given as 1925, the description on the back would place it prior to 1913 when the direct DC motor drive was abandoned in favour of cables and winches.

In 1947 management of the bridge passed to the short-lived British Transport Commission as part of the Labour administration's transport nationalisation programme, but the BTC was scrapped in 1962 after just 15 years when Harold Macmillan's Conservative government passed their Transport Act – by which time the transporter bridge already been closed for a year and was in the early stages of being dismantled.

With the rapid increase in road traffic, the bridge's shortcomings as a primary transport system had already become very apparent. Waiting times for both foot passengers and vehicles had been getting longer and longer since before the end of the Second World War, and

The surviving sandstone building, later used as an electricity substation, originally housed waiting rooms and a ticket office.

Above: On the Widnes shore, the paved approach to the transporter bridge survives. Beyond the end of the ramp the roadway extended out across a cast iron pier to meet the north towers.

Left: At water level, a few courses of dressed stonework remain from the foundations of the docking pier.

RC 2   RUNCORN. THE TRANSPORTER BRIDGE   Photo by J. F. LAWRENCE

A view of the bridge and Manchester Ship Canal from the Runcorn side, by J. F. Lawrence, published by Mason's 'Alpha Series' postcards in the late 1940s.

the associated congestion on both sides of the bridge had been continually increasing.

The decision to replace the transporter with a fixed bridge was inevitable, and planning work started on a steel through-arch bridge in 1956, crossing the river and the Ship Canal at the narrowest point between the transporter bridge and the 1868 railway bridge. It took nearly five years to build, eventually being opened to traffic in July 1961 by HRH Princess Alexandra.

As traffic continued to increase, that bridge was widened between 1975 and 1977, being named the Silver Jubilee Bridge on completion, but by the early years of this century, even its capacity was insufficient, and in 2017 it was joined by the new £600M three-pier, six-lane, cable-stayed Mersey Gateway crossing. Such is the cost of coping with increased traffic flow.

For travellers, the transition from the transporter bridge to the fixed high level bridge meant that in 1961 crossing the river had become free of charge for the first time; but with the opening of the latest bridge, and amid much protest from local users, even using the 1961 bridge now involves the payment of tolls.

The last crossing of the transporter bridge was made on 22 July 1961, with demolition starting almost immediately thereafter. As with most things as closure looms ever closer, the last few days saw the bridge more heavily used than it had been for years, with hundreds of people keen to say 'they were there' before it was gone forever.

Ironically, the cost of the two-year demolition programme far exceeded the £130,000 cost of the bridge's five-year construction programme, completed 56 years earlier.

It is a considerable loss to engineering history that the longest and most impressive transporter bridge ever built should have come to such an ignominious end, but its sheer scale would have been the biggest obstacle to any idea that it might be worth preserving – not that serious consideration was ever given to such an idea back in 1961. It had simply outlived its usefulness, overtaken by the post-war transport revolution.

What little evidence of its existance survives today gives no clue to its ground-breaking engineering – a couple of notice boards give a brief account of its history, but that is all. The bridge's operational costs were considerable even back then, and much higher than either of the two of Britain's three surviving transporter bridges which now enjoy considerable popularity – primarily as tourist destinations – in Newport and Middlesbrough.

Surely something more could be done with the site today – a series of information panels in the derelict waiting rooms would be a welcome addition for visitors fascinated by the bridge.

In the meantime, all that remains of Britain's longest transporter bridge today is the approach ramp and its associated buildings on the Widnes side, part of the approaches at Runcorn and, of course, the photographs and memorabilia produced during its lifetime.

The gondola makes its last-ever crossing, 22 July 1961, marking a little over 56 years of service.

· Pont Gludo · Transporter Bridge · Pont Gludo · Transporter Bridge ·

Day Visitor:
Includes unlimited crossings and
access to the High Level Walkway

Ymwelydd Dydd:
Yn cynnwys nifer di-derfyn o groesiadau
a mynediad at y Dramwyfa Lefel Uchel

Opening Hours/Oriau Agor
Opening hours vary, please see website for details

www.newport.gov.uk/transporterbridge

Adult/Oedolyn £3
01876

NEWPORT
TRANSPORTER BRIDGE
25658
PLATFORM 6d

Williamson, Ticket Printer. Ashton

BRIDGING THE USK AT NEWPORT

Seen from a distance the Newport Transporter Bridge, which spans the River Usk in South Wales, appears to be a lightweight delicate structure, but up close, the complexity of the engineering is a revelation. Delicate it most certainly is not, its robustness underlined by the fact that it has stood firm against the elements for more than 110 years.

The longest transporter bridge still standing in the world, the Newport bridge – 'Pont Gludo Casnewydd' in Welsh – is the only example of a Ferdinand Arnodin design ever built outside France.

There had long been a need for a bridge across the river at around this point, but there were considerable logistical problems associated with building a conventional crossing. Newport was a busy and expanding port, with much of

Opposite: The suspension cables seen from the walkway across the top of the stiffening girder. The 16 main cables weigh a total of 199 tons and are embedded in masonry anchorages 137 metres back from the towers.

Insets: Tickets from the present day and the 1930s.

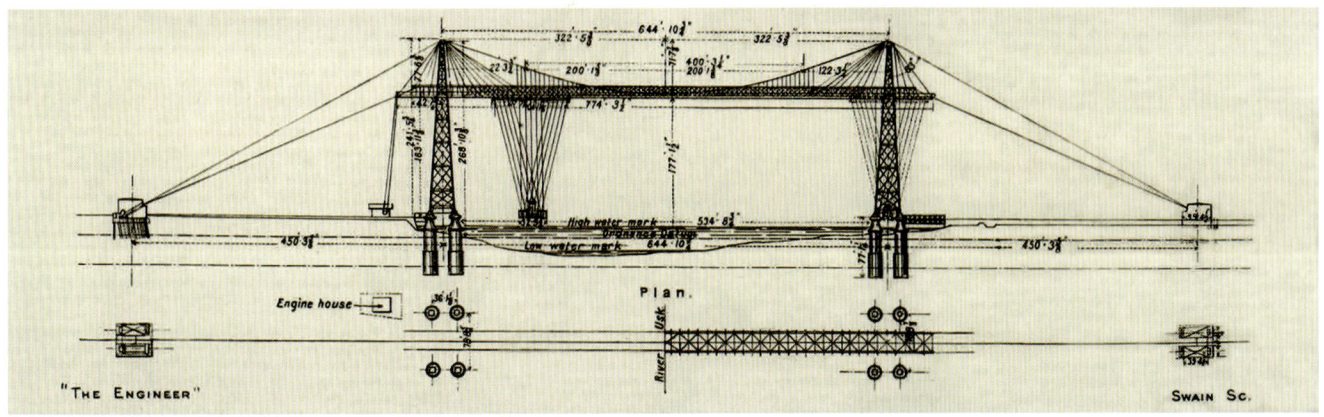

Above: The plan of the bridge illustrated in the journal *The Engineer* on 14 September 1906. The gondola is shown twice – once mid-stream, and also at the west end, docked between the towers.

Left: From a postcard published in 1906.

TRANSPORTER BRIDGE, NEWPORT, MON

the shipping passing further up river to the docks – and in the closing years of the nineteenth century when the project was first mooted, tall masts required considerable clearance beneath any bridge which might be built. The flatness of the surrounding countryside also worked against those who would have a conventional crossing.

With industries expanding on both sides of the river, the town centre bridge was just too far away – a four mile walk was needed to get from one side of the river to the other at the point where the transporter bridge now stands. The only alternative, a small ferry with very limited passenger capacity, also had operational problems due to the considerable rise and fall of the tides at this point on the river – up to 8 metres at times.

Industrial expansion on the east side of the river was well under way before the end of the nineteenth century, and amongst the those pressing for easier access across the Usk for their workforce was William Royse Lysaght, whose company, John Lysaght Ltd., had just recently relocated their steel rolling business from Wolverhampton. In 1898 they were in the process of building new rolling mills at their Orb Ironworks on the Pill Farm site east of the river.

That a transporter bridge was even considered was down to Newport's Borough Engineer, Robert Haynes, who had heard of Ferdinand Arnodin's bridges being built on the Continent and encouraged councillors and planning officials to go to France and take a look for themselves. Just as with Thomas McGilvray and the

Above left: The completed beam, published in *The Engineer*. The travelling frame wheels can be seen in the bottom right corner of the photograph.

Above right: 110 years later the girder has developed a bit of a sag, but the bridge still operates smoothly.

Opposite left and right: These photographs of the construction of the stiffening girder were published in *The Engineer* on 14 September 1906.

Opposite below: A 1905 view taken from the wharf near to the old dock entrance before the travelling frame and gondola were installed.

The article in *The Engineer* included numerous detailed drawings showing the design and construction of various parts of the bridge.

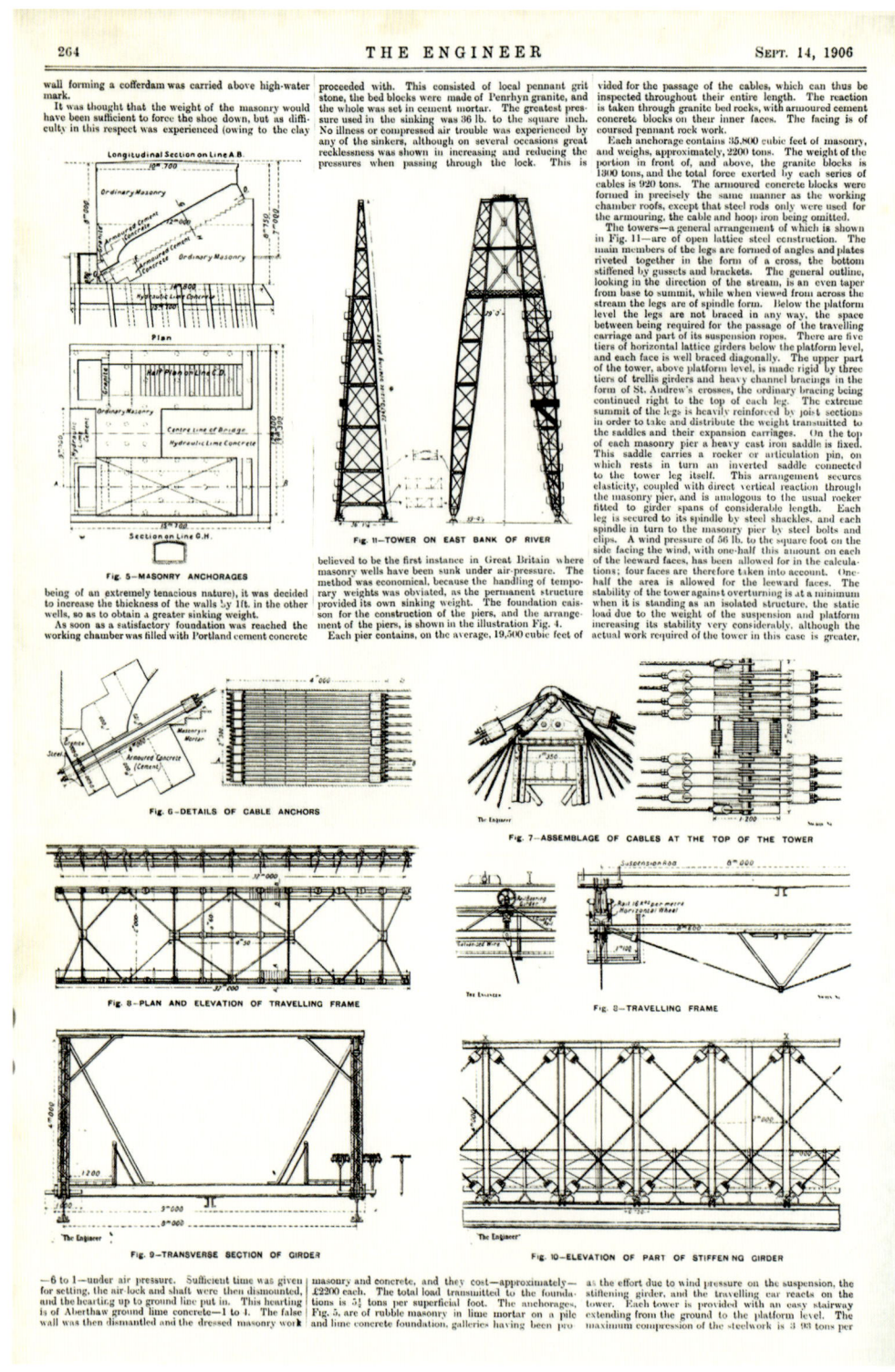

city officials from Duluth in Minnesota, their journey took them to Rouen just weeks after Arnodin's bridge over the Seine had opened in 1899.

Whereas the American bridge which eventually opened across the harbour mouth at Duluth in 1905 – the year before completion of the Newport bridge – looked radically different from Arnodin's designs, the Newport bridge was a

Above: The foundation piers on the Newport bridge – their bell-shaped design would be replicated on several other Arnodin transporter bridges.

Left: This early photographic postcard, published around the time of the bridge's opening in 1906, shows clearly the position of the winch-house at the east end of the bridge.

The combination of suspension cables and cable stays, pioneered on the Vizcaya bridge, was a feature typical of most of Arnodin's early transporter bridges.

logical evolution of his ideas, bearing all the hallmarks of his early twentieth century French bridges.

A three-page article on the genesis, design and construction of the bridge was published in the 14 September 1906 issue of the journal *The Engineer* which began by discussing an 1890 plan to build a pedestrian subway under the Usk – a plan which came to nothing.

'Six years later—in 1896—the subway scheme again came under review, and the suggestion was made that a "transporter" type of bridge—an example of which had recently been erected at Bilbao—was the most

The 1893-built paddle-steamer PS *Albion* passing beneath the bridge c.1910.

One of many postcards of passengers on board the gondola, with just a simple chain to stop them ending up in the river.

Right and below: The bridge can either be controlled from the winch-house on the east bank, or from the driving cabin above the gondola as here. At low tide, the size of the stone piers on which the towers are fixed can be appreciated. The supporting cables which can be seen running to ground beyond the towers are embedded into 2,235 metric ton masonry-block anchor chambers more than 135 metres back inland from each tower. The slender eliptical profile of the towers is a fine example of how function and elegance can be combined in an effective design.

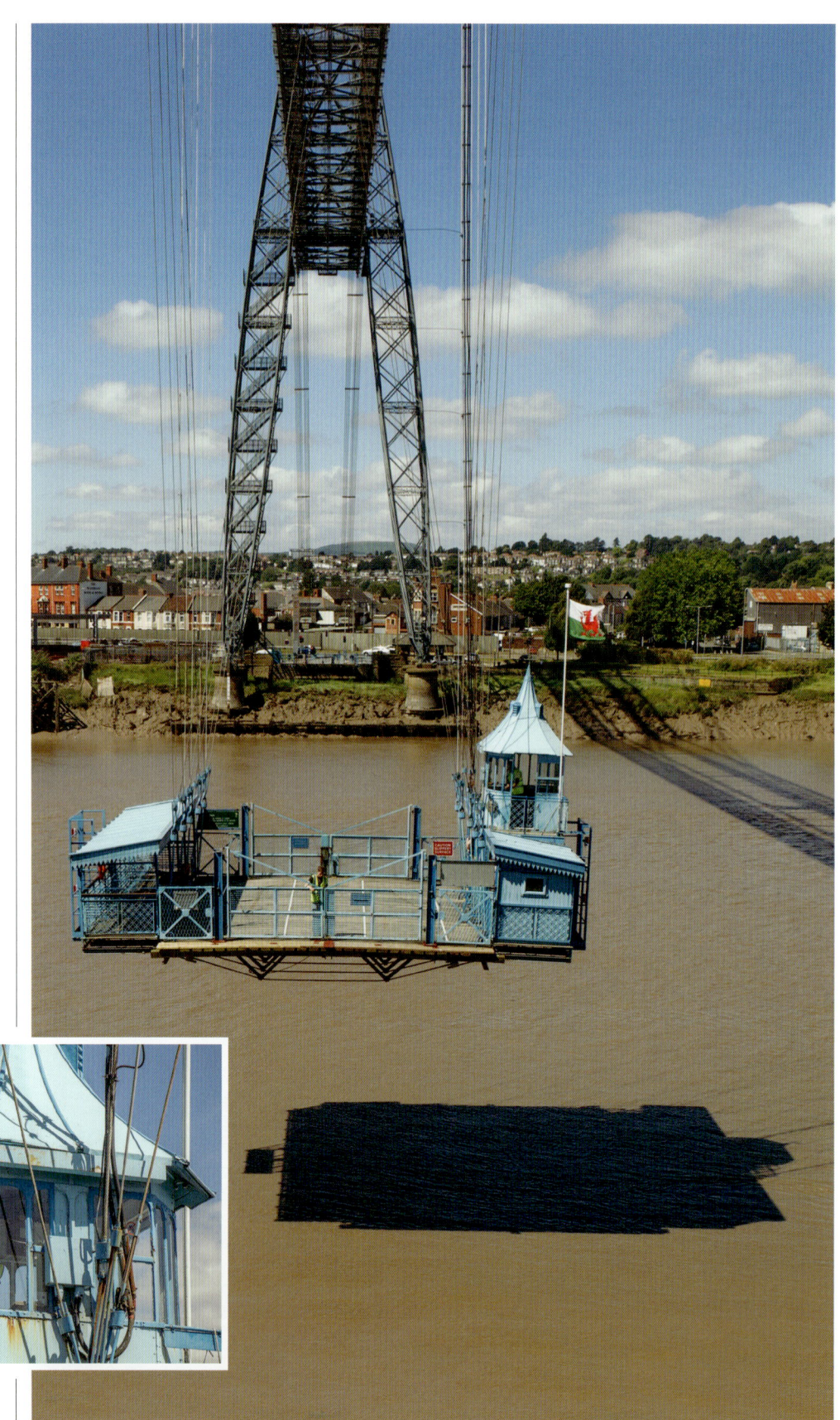

satisfactory solution to the problem. The idea was at that time somewhat too advanced for local opinion, and the question remained in abeyance for some short time. Two years later—in 1898—owing to the development of the town on the eastern side of the river, the question of traffic facilities grew acute, and it became necessary that a definite scheme should be proceeded with at an early date. Schemes of all kinds were prepared and considered, including ferries, high level, swing, lifting, bascule, and rolling bridges and subways, but the choice ultimately fell upon the "transporter" type of bridge. The scheme for a subway of sufficient capacity to deal with vehicular traffic was abandoned on the score of cost; high-level bridges with sloping approaches were out of the question for financial reasons; swing bridges were rejected owing to the obstruction they offer to the free use of the waterway, and for the additional reason that complications were not unlikely to arise owing to the reduction in the section of the channel due to the large piers required. It is a fact which is not generally appreciated that swing bridges of limited width can only be navigated with safety against the current, a condition which cannot be observed in the river Usk, with its relatively short period of high tide, during which the movement of shipping is alone possible. The "transporter" type of bridge was favoured on account of the little or no obstruction which would be offered to the shipping, whilst the cost of erection was well within the resources of the community. Detailed schemes were prepared, and in 1899, after having inspected the Rouen "transporter", a practically unanimous decision was arrived at by the governing authorities at Newport in favour of that type, and the proposal received parliamentary sanction in 1900. Detailed calculations were made, designs prepared, and contracts for the necessary work entered into by August 1902, since which date the work has proceeded apace, the bridge being formally opened for traffic by Viscount Tredegar on Wednesday last.'

The justification for a transporter bridge, so clearly expressed in the article, was echoed in the development of most of the other transporter bridges built – lower construction cost, short construction time and minimal invasion of the shipping channel being crucial in so many cases.

By the time it opened, it became the world's ninth transporter bridge, and the sixth to be designed by Arnodin, but unlike his others – which were mostly built at his foundry at Chateauneuf-sur-Loire – the Newport bridge was engineered and constructed by a consortium of UK companies. Specialist bridge builders, Alfred Thorne & Sons of 7 Carter Street, Westminster, were the main contractor and the steelwork was manufactured by the Cleveland Bridge & Engineering Company.

The steel cables were made by W.B. Brown & Company at their Globe Works in Liverpool, and the electrical equipment – two 35bhp (26kw) DC electric motors to drive the winches – were supplied by Robert W. Blackwell & Company, electrical manufacturers, engineers and contractors, of Victoria Street in Westminster.

Once just a chain was used at either end of the gondola, but in today's safety-conscious environment, securely locked gates are considered necessary to perform the same purpose.

While they supplied the motors and winches, they apparently did not manufacture them. The motors were built by the Lancashire Dynamo & Motor Company Ltd., who had established their vast engineering works on what was then the brand-new Trafford Park Industrial Estate adjacent to the Manchester Ship Canal and Docks in 1899.

Arnodin supervised the project through his associate and agent Georges Camille Imbault, a native of Chateauneuf-sur-Loire, who had gone to work for Arnodin at an early age, working with him on many projects before moving to London in 1901, apparently to learn English, which he saw as a means of broadening his career opportunities.

That decision proved to have been a very well informed one, and his newly acquired language skills saw him working on major projects in several parts of the British Empire.

Arnodin seems to have taken on the design work for the Newport bridge in early 1901, and by the following year he had retained Imbault as his representative on the project.

Imbault's association with the Cleveland Bridge & Engineering Company came about in the following year at around the time that the company was awarded the contract for the bridge, thus providing an essential link between the designer and the builders.

It would prove an enduring and successful partnership, with Imbault working on many important projects for the company – most notably acting as Engineer on the 1903 Victoria Falls bridge over the Zambesi, and as Consulting Engineer on both the Tees Transporter Bridge between 1908 and 1911 and the Sydney Harbour Bridge which opened in 1930.

Work started on site in 1904 and took just under two years. The first and major operation was the construction of massive steel and concrete caissons – watertight chambers – in which workmen would eventually excavate deep down on to the bedrock. The procedure was explained in *The Engineer*:

> 'The first 12ft. or 14ft. was done by a grab and by men working in the open, the air lock was then mounted, and sinking under pressure proceeded with, the walling being carried up simultaneously so that at all times the top of the well was above high water.'

Sinking the caissons proved less straightforward than had been anticipated as, despite the huge weight of steel and concrete from which they were made, they did not easily sink down as planned as the excavations continued – the resistance of the clay-heavy soil proved to be much greater than the engineers' calculations had predicted.

Arnodin's construction methodology is evident in this view taken from the top of the Newport tower at low tide. The bell-topped masonry piers – of near-identical design to that used in several of his French bridges – can clearly be seen standing on their concrete-filled caissons and protected by wooden shuttering.

The problem was relatively simply solved, however, by increasing the thickness of the walls of each caisson by one foot, thus significantly increasing their weight and overcoming the soil's resistance.

Once the excavations had got down as far as the bedrock – and that meant more than 24 metres below the spring high water level on the west bank and 26 metres on the east – the caissons were then partly filled with concrete to form solid bases on to which were built four of Arnodin's signature bell-topped masonry piers on each side of the river. These formed the bases on which the partially pre-fabricated steel towers were subsequently erected.

Each completed foundation contained more than 550 cubic metres of masonry and concrete, and each cost £2,200 – a relatively small sum even in 1906 when considered within an expenditure estimated at £98.753 for the project as a whole.

As *The Engineer* pointed out, some of the cost-saving techniques used for sinking the foundations had involved pioneering engineering technology.

'The greatest pressure used in the sinking was 36lb. to the square inch. No illness or compressed air trouble was experienced by any of the sinkers, although on several occasions great recklessness was shown in increasing and reducing the pressures when passing through the lock. This is believed to be the first instance in Great Britain where masonry wells have been sunk under air-pressure. The method was economical, because the handling of temporary weights was obviated, as the permanent structure provided its own sinking weight.'

Each completed tower stood 74 metres high and was made up of 277 tons of steel. With the additional weight of the cables, iron rods, the gondola, its overhead carriage and whatever weight of passengers and vehicles the gondola was carrying, those foundations were – and are – constantly under considerable loading.

Opposite: The underside of the main beam as the wooden-clad travelling frame starts its journey across the river from west to east. It was in order to facilitate industrial development on the east side of the river that the transporter bridge project was initially considered. The stiffening beam is made up of two riveted latticework girders stiffened with vertical members every two metres and cross-braced diagonally with steel cables. The length of the carriage or 'traveller' is three times that of the gondola, the widely-splayed cables from which the gondola is slung helping minimise any tendency to rock or sway in windy conditions. The winch-house can be seen between the far towers.

The artist has got the geometry of the bridge all wrong in this postcard c.1905, with the gondola riding several metres too high above both the water and its docking station.

This card was produced by local Newport publishers Greenland & Sons in the closing weeks of 1904, long before work on the bridge was complete. The un-named artist has omitted the winch-house – or probably it had yet to be built when the sketch was made.

Arnodin was already embracing several other 'new technologies' – the compressed air system was obviously one, and the use of electric cranes and winches during the construction – powered directly from Newport's own municipal electricity supply – both cut costs and speeded up assembly.

'An electrically-driven crane was used for the erection work, and proved most economical, as less than 100 units, costing under £8.10s., were ample to lift and place in position the whole of the temporary scaffolding and staging, as well as the permanent structural steelwork in both towers.'

For his French projects, Arnodin had evolved a tried and tested production-line approach to the manufacture of standard components in his Loiret foundry, but for the Newport bridge the fabrication was being undertaken not in Chateauneuf-sur-Loire, but in the various factories to whom the work had been sub-contracted by the Cleveland Bridge & Engineering Company.

According to a number of subsequent accounts, that had led to some confusion over measurements, with Arnodin working in metric units while drawing up his plans and engineering drawings, and Alfred Thorne & Sons other sub-contractors working in imperial measures.

The exact nature of any such confusion – had it ever actually occurred – has, however, been lost in the mists of time and the story may well be apocryphal as several of the measurements are clearly imperial. The travelling frame, for example, is exactly 104 feet long – 31.7 metres.

Had it originally been designed in metric measure, it would more likely have been 32 metres, which would equate with an imperial measure of 105 feet. Similarly, the gondola measures 40 feet by 33 feet, and while 33 feet is

Above left: The base of the south-east tower showing its riveted construction. The towers are fixed to the piers using articulation pins in cast-iron saddles to allow for movement. The signpost records the distances to the world's seven other surviving transporter bridges.

Above right: Cables from the winch-house run on guide wheels the full length of the beam.

a neat 10 metres, its 40 feet width uncomfortably converts to 12.2 metres. If any confusion did arise, however, it must have helped having Imbault, with his knowledge of Arnodin's working principles, on hand to resolve such issues.

The measurements quoted in the article in *The Engineer*, however, do show some confusion with a mixture of both imperial and metric, neither of which leads to round figure conversions either way. But that, of course, is down to careless journalism and cannot be blamed on either Arnodin or Imbault:

'The suspension cables are sixteen in number—four inside and four outside each of the stiffening trusses. Each cable is composed of 127 wires 21.7mm in area, the section of each cable being 4.273 square inches'

The cable winch motor and cable. The winch-house sits on an elevated platform behind the east towers.

The drive cable goes up to the stiffening beam and the gondola carriage – referred to on the plans as the travelling frame, the continuous steel cable running the full width of the bridge and back.

Some of the 60 wheels on the travelling frame.

The effectiveness of the bridge owes much to the pioneering design work of Alberto Palacio as well as to Arnodin's engineering skills. Right from the construction of the first transporter bridge at Bilbao, the two men had recognised the value of supporting the stiffening beams with a hybrid of vertical iron rods hung from the main suspension cables, and wire cable stays radiating from the tops of the towers.

Stress calculations had shown them that the cable stays would take the bulk of the loading as the gondola moved towards either tower, the stresses being shared with the suspension wires when the gondola was mid-stream.

Below left: The friction drive linkage between the electric motor and the winch, with the main brake pad below.

Below right: The winch cable goes up to the end of the beam, through an aperture in the winch-house roof.

The Engineer reckoned that under maximum loading, the tension on the main parabolic cables was 1.93 tons per square centimetre (12.27 tons per square inch), while the tension on the cable stays measured 2 tons per square centimetre (12.7 tons per square inch) – thus achieving a near perfect dispersion of the loads.

To counterbalance the whole structure, the 16 main suspension cables –with a total weight of 196 tons – were anchored deep in concrete and masonry bunkers around 140 metres back from each tower. The radiating cable stays added another 19 tons to the weight.

Each of the main cables was formed as a continuous length from anchorage to anchorage, moving on roller saddles as they passed over the top of each tower to allow for expansion and contraction, which occurred as temperatures and loadings changed. The anchorages each contained more than 1000 cubic metres of masonry and concrete and each weighed in excess of than 2,200 tons.

Two 35bhp 25kw DC electric motors wind and unwind around 500 metres of winch cable – known as the long and short 'ropes' – to haul the travelling frame and the fully-laden gondola suspended below it across the river. The winches are therefore hauling a total combined weight of 50.7 tons.

That it is still the original motors and winches which do the job today says a great deal about the quality of equipment sourced for the bridge by R.W. Blackwell Limited of Westminster.

Walking the main beam of the Newport Transporter Bridge, with the streets of the South Wales city 55 metres below. The winch cable can be seen in the foreground.

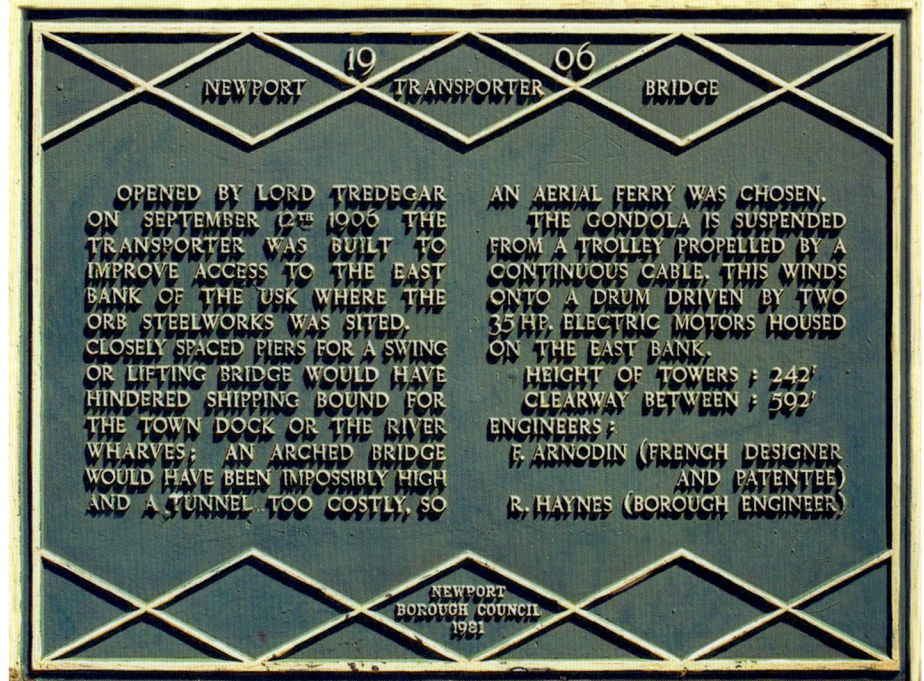

Above left: The 1906 maker's plate on the winch motors which still drive the 112-year-old bridge.

Above right: Based on a painting published as a postcard in late 1904, this jug was probably produced to commemorate the opening of the bridge in 1906.

Left: This plaque was unveiled on the bridge's 75th anniversary.

The winches – which Arnodin originally intended to be mounted on top of the gondola as with his Rouen bridge but changed his mind – are turned by friction drive rather than toothed cogs. This was felt less likely to incur damage if the bridge had to be stopped quickly for any reason.

According to today's engineers, damage to a cog-driven system could have been caused by something as simple as a sudden gust of wind briefly impeding the passage of the gondola and carriage. Given the winds around Newport, while friction drive can cause occasional slippage, it has clearly brought enduring benefits throughout the bridge's 112 years of operation, as the winch and motors are the originals fitted in 1906.

FINALLY CROSSING THE TEES

Thirty-eight years after Charles Smith first proposed building a transporter bridge across the River Tees at Middlesbrough, just such a bridge was opened with great ceremony by Prince Arthur of Connaught. It even looked a bit like Smith's and, affectionately known as the 'Tranny', it is still doing what it was designed to do more than a century ago.

It was designed by the Cleveland Bridge & Engineering Company of Darlington, but built by Sir William Arrol & Company of Dalmarnock, Glasgow, who had also built Scotland's iconic Forth Bridge. Construction took just over two years.

Planning for the bridge began in 1906 when Middlesbrough Corporation held exploratory meetings with Ferdinand Arnodin, William Edwin Pease, Managing Director of the Cleveland Bridge & Engineering Company and its chief engineer Georges Camille Imbault. Imbault and Arnodin had worked together on a number of Arnodin's bridges – in Tunisia, Rouen, and Newport, so there clearly existed a great deal of trust and understanding between the two men. The presence of Arnodin in a consultative capacity was a wise precaution given the many patents he had been granted which gave him almost unlimited rights over transporter bridges.

A 1907 Act of Parliament had authorised construction of a transporter bridge linking Middlesbrough and Port Clarence and permitting the withdrawal of the long-established ferry service across the river once the new bridge was opened. Imbault and Cleveland Bridge were appointed to design the bridge.

Opposite: Still a well-used transport link – albeit operating only six days a week – the 'Tranny' has now been a feature of the Middlesbrough skyline for more than a century.

Below left: The Tees Transporter bridge featured on a stamp in 2015 as part of a Royal Mail series on iconic bridges.

Below right: One of the illustrations of the bridge photographed for the official booklet published at the time of the opening.

Right: Dated 3 June 1910, this photograph is captioned 'Completed Wing Wall on the Middlesbrough side' and shows work underway on the foundations for the bridge. It is from a series of progress images taken by Robert Compton Clifford whose studio was at 18 Wilson Street, Middlesbrough.

Below: In another of Clifford's photographs, the towers on the Port Clarence side are starting to take shape – this image was taken 1 August 1910 – with the soon-to-be-redundant ferry *Erimus* steaming alongside.

Construction started in late 1909 and the bridge was opened on 17 October 1911 by Prince Arthur, with Sir William Arrol himself reportedly attending the ceremony.

Throughout the bridge's lifetime, the gondolas have already crossed the river more than six million times, a total distance travelled of more than 27 times round the world.

So iconic is the structure that locals are understandably very protective of it, some perhaps slightly too protective. When Dick Clements and Ian le Frenais wrote a story about it being dismantled and sold to Native Americans – in the third television series of *Auf Wiedersehen, Pet* in 2002 – the tv special effects people did such a realistic job of apparently showing the bridge being dismantled that there were those who thought it really had been moved and sold, and the BBC had to publish on-screen reassurances that it was still *in situ* and hadn't been touched.

Since then, the bridge has undergone a major £4M overhaul, a new gondola has been constructed and installed – the third in the bridge's lifetime – and since 2015 the 'Tranny' has been back in service doing what it was built to do over a century ago.

Charles Smith's 1873 idea of a 'ferry bridge', or transporter bridge over the Tees had been resurrected in 1901 by Alderman Joseph McLauchlan who had reportedly visited both Palacio and Arnodin's Viscaya Bridge in Portugalete and Arnodin's Rouen bridge, opened in 1893 and 1899 respectively. He had been advised, perhaps a little unrealistically, by the engineer Charles H. Gadsby whom he had consulted, that a transporter bridge along similar lines to Smith's design could be built for around £40,000 – just one third more than Smith had estimated 28 years earlier.

Gadsby and Arnodin had jointly published a proposal to build the Shields Transporter Bridge across the Tyne. That proposal, however, was clear evidence

The journal *The Engineer* published this elevation and plan in its issue of 29 September 1911.

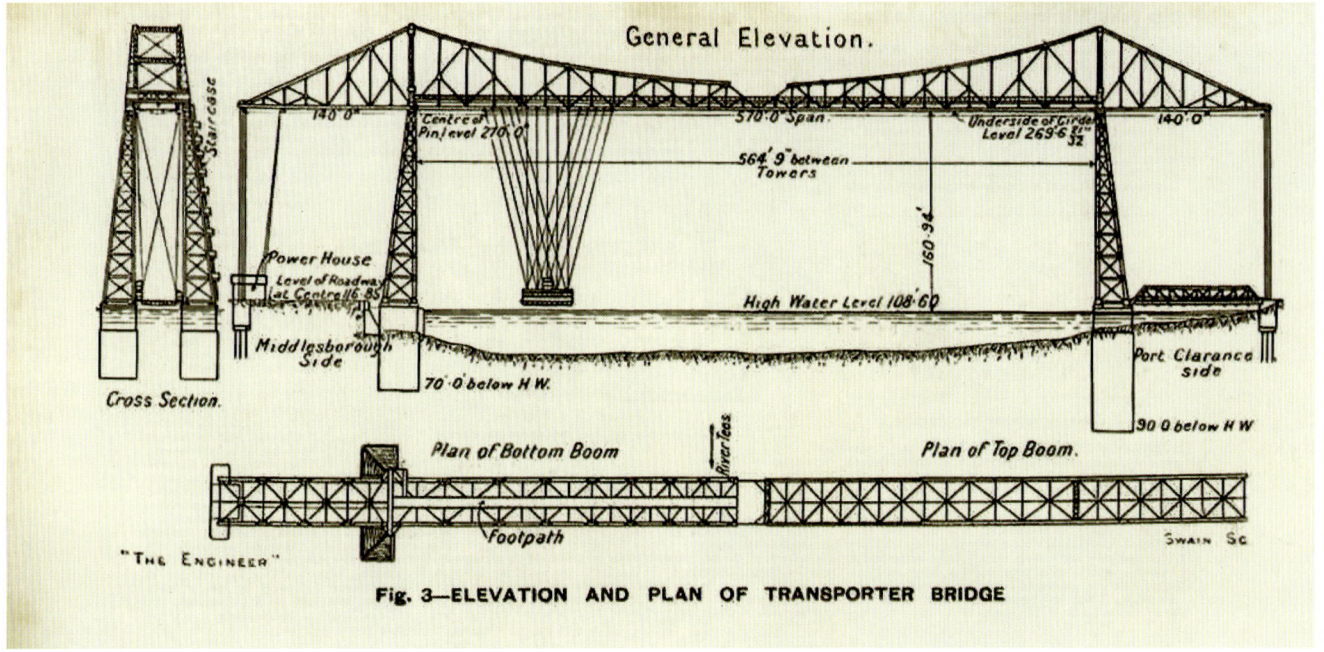

Fig. 3—ELEVATION AND PLAN OF TRANSPORTER BRIDGE

336   THE ENGINEER   SEPT. 29, 1911

NEW TRANSPORTER BRIDGE AT MIDDLESBROUGH

DROPS, SHIPPING, AND THE FIRST FERRY ON THE TEES, 1841

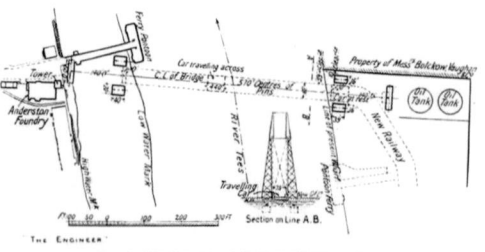

PORT CLARENCE IN 1841

THE TRANSPORTER BRIDGE OVER THE RIVER TEES.

THE transporter bridge over the river Tees, connecting Middlesbrough and Port Clarence, Fig. 1, which the County Borough of Middlesbrough has had under construction during the past two years, will be formally opened for traffic by His Royal Highness Prince Arthur of Connaught, K.G., on Tuesday, October 17th. The bridge is of outstanding 'importance not merely by reason of its constructional features, but also by the prominent part which it is destined to play in the future industrial development of Teesside, and more especially of the large area on the north bank of the river immediately opposite

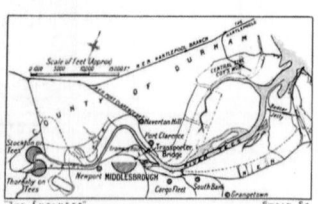

Fig. 1—SKETCH MAP OF THE DISTRICT

to the town of Middlesbrough. For many years past the increasing traffic across the river Tees at Middlesbrough has caused attention to be prominently called to the necessity of providing some better means of transit than that afforded by the existing small ferry boats, and a number of rival schemes have been brought forward for attaining the desired end. Thus there have been advocated a tunnel under the river, swing bridges, a lifting bridge, and lastly the "bridge ferry," or transporter bridge, of which it is now more especially our object to speak. Before, however, passing to a detailed description of this new bridge, it may not be without interest to place on record a short review of the history of the traffic facilities

across the river Tees, and the various negotiations which took place preliminary to the construction of the bridge.

The origin and position of the ferry from Middlesbrough to Port Clarence is somewhat difficult to trace, although it is conjectured that some primitive means of crossing the Tees was in vogue by the monks who inhabited the ancient cell dedicated to St. Hilda in their communications with their brethren at Hartlepool, but at a point a little higher up the river than the present site. In ancient times a ford is also believed to have been in existence across the river opposite Newport, and in

Fig. 2—POSITION OF THE TRANSPORTER BRIDGE

the absence of any reliable data supporting this view there is a boulder in the Albert Park, Middlesbrough, which was dredged from the bed of the Tees by one of the Tees Conservancy Commissioners' dredgers a few years ago, and which is inscribed as one of the stepping stones of such a ford. Certain it is that, with the advent of the coal export trade in this district, a considerable traffic sprung up in the thirties between the two banks of the river and caused a row-ferry to become an accomplished fact. It was in 1830 that the Darlington and Stockton Railway was continued from Stockton to

Middlesbrough, where the shipment of coal from staithes grew apace, and was at that time the staple and practically the only trade of the town. This port was then designated "Port Darlington." In 1832 the railway of the Hartlepool Harbour and Railway Company was extended to Samphire Bats—now called Port Clarence—where coal staithes were also erected. About this time a horse coach ran from Port Clarence to Billingham, where passengers *en route* to Hartlepool, Durham, Newcastle, &c., joined the trains. In the years 1838 and 1839 two attempts at establishing river, passenger, and goods-carrying traffic between Stockton, Haverton Hill, Middlesbrough, and Port Clarence were attended by failure to the Stockton and Port Clarence Steam Packet Company. The ferry continued as a private enterprise up to 1856, when powers were obtained in a local Act enabling the Middlesbrough Corporation to establish a public wharf and passage over the Tees. Thereupon the corporation acquired from the Middlesbrough owners the ferry site on the south side of the river, and combined with the West Hartlepool Harbour and Railway Company, which held the interests on the north side, for the ferry to be run as a joint concern. Complaints were, however, rife at the delay of the company in carrying out its part of the agreement, and in 1857 the Middlesbrough Corporation was successful in obtaining parliamentary powers to construct a ferry landing on the Durham shore, and the following year it obtained the full control of the ferry. In 1859 a new boat was purchased for £20, which was considered a more convenient and safe boat for the ferry! In 1860 the question of providing steamboats for the ferry service came under consideration, and two years later a steamboat, The Progress, was built by Mr. Frederick, and launched from his shipyard at Middlesbrough. She was

of Arnodin's continuing involvement with the idea of building transporters in Britain – the 1906 Newport bridge – his only British bridge – was, of course, still to be built.

He was closely involved in the whole debate about how best to bridge the busy Tyne, and through his agent and former colleague George Camille Imbault – who by then was employed by the Cleveland Bridge & Engineering Company as their Chief Engineer – he would retain a close involvement in the eventual construction of the bridge which we see today.

It would, remarkably, be a further five years, however, before any real progress towards the design and construction of the bridge was made, and more than two further years before it was complete.

In 1911, the journal *The Engineer* celebrated the imminent opening of the long-awaited bridge in its issue of 29 September, with a two-page account of its history and construction.

'The bridge is of outstanding importance not merely by reason of its constructional features, but also by the prominent part which it is destined to play in the future industrial development of Teesside, and more especially of

Opposite: An account of the bridge's genesis and construction, as reported by the journal *The Engineer* on 29 September 1911.

Left: One of Clifford's images from 11 March 1911 with the cantilever beam starting to take shape on top of the Middlesbrough towers. This image gives a clear view of the temporary scaffolding tower which helped support the weight of the beam during construction.

the large area on the north bank of the river immediately opposite to the town of Middlesbrough. For many years past the increasing traffic across the river Tees at Middlesbrough has caused attention to be prominently called to the necessity of providing some better means of transit than that afforded by the existing small ferry boats, and a number of rival schemes have been brought forward for attaining the desired end. Thus there have been advocated a tunnel under the river, swing bridges, a lifting bridge, and lastly the "bridge ferry", or transporter bridge of which it is now more especially our object to speak.'

There then followed a brief account of the history of traffic across the river, and of Charles Smith's 1873 bridge proposal, but more interesting remarks followed:

'In November 1905, notice was served upon the Middlesbrough Corporation of an intended application to the Light Railway Commissioners for an order authorising the construction of a light railway between Middlesbrough and West Hartlepool, this scheme including the building of a transporter bridge over the Tees so as to link up the existing tramways with the proposed light railway to West Hartlepool. The tramways in Middlesbrough form part of the Imperial Tramways Company, Limited, obtained by Provisional Order in 1897, and the Middlesbrough Corporation viewed with disapproval any scheme which further diminished its control over its roads.'

Middlesbrough's steam-powered trams were, at the time that this new proposal was made public, being converted to electric running by Imperial's subsidiary the Middlesbrough, Stockton and Thornaby Electric Tramways Company, and the parent company was buying up tram systems nationwide and seeking new

Just three months later, with construction nearing completion, Clifford took this photograph on 16 June 1911. The builders were just a few weeks away from removing the temporary scaffolding.

ways to extend its networks. The Middlesbrough & West Hartlepool Light Railway proposal, had it been built, would have been a perfect fit.

According to a report in the *Northern Echo* on 17 February 1906, the project was the brainchild of Charles Watson of Birkenhead and Charles Straker of Newcastle. The total cost of the tramway was initially estimated at just over £120,000, of which £69,430 would be the cost of building the bridge and its approaches. The cost of the gondola – referred to as the 'conveyor bridge car' – was estimated at £1,500. Interestingly, in the proposal documents, the bridge was sometimes referred to as a transporter bridge, and other times as a 'conveyor bridge'.

The official 1911 souvenir booklet recounted the story of both bridge and port.

The gondola photographed shortly after its installation.

One of the postcards produced at the time the bridge opened included a picture of Alderman Jospeh McLaughlin, one of its strongest advocates.

An early tinted postcard of the bridge.

Transporter Bridge, Middlesbrough.   N° 2.

Transporter Bridge, Middlesbrough.   N° 3.

Left and below: Two early postcards from a series showing the hugely crowded gondola nearing the Middlesbrough shore, and arriving at the jetty.

By the time the second bridge was proposed, industrial links between Bilbao and Middlesbrough were well established, so the success of Palacio's Viscaya Bridge would have been known to Teesside industrialists who already had investments in Bilbao's ironworks. Indeed there was already a thriving trade between the two ports, with large quantities of both iron ore and pig iron being shipped from the Spanish port to the Tees.

Had it been built, it would have joined the Duluth bridge in the United States as being one of only two transporter bridges in the world designed to transport tramcars. Instead, it joined the list of other proposed tramway transporter bridges which never got beyond the drawing board – across the

New Transporter Bridge, Middlesbrough.

In this October 1911 postcard, the 1884-built former ferry *Hugh Bell* has already been withdrawn and beached near the bridge while another, perhaps the 1887-built *Erimus*, is seen mid-stream.

Ribble in Lancashire, across the entrance to Poole Harbour in Dorset, and between Portsea and Hayling Island in Hampshire.

Railway tracks would eventually be fitted to a transporter bridge in 1916 – the last to be built in Britain – when Joseph Crosfield & Sons of Warrington completed construction of their second private transporter over the River Mersey.

Interest in the proposal was considerable, as the ferry service was as incapable of coping with traffic demand as Charles Smith had suggested it would be more than 30 years earlier.

As a financial inducement, the Middlesbrough & West Hartlepool Light Railway offered the council 10 per cent of the gross income from their bridge with the council having the right to buy the bridge outright after 35 years – at the same price as its construction costs.

Further financial inducements were offered, all to no avail – one reason apparently being the company's claim that the bridge would cost £80,000 to build, while councillors had estimated the total cost at £50,000.

At the same time that Watson and Straker were commissioning plans for their tramway bridge, Ferdinand Arnodin and Charles H. Gadsby were seeking support from Middlesbrough Council for their own transporter bridge design. No illustrations have been located during the research for this book of either their proposed bridge or Watson and Straker's – which was to have been designed by George Griffin Eady and Alfred S. Frech and may have been based on Smith's 1873 design. It seems probable, however, that Arnodin and Gadsby's proposal, while crossing the river at about the same point as the present bridge, would have utilised Arnodin's now 'standard' design of a suspension bridge with cable stays. Both proposals, like Smith's, were rejected.

The threat posed by the tramway proposal, however, certainly galvanised Middlebrough Corporation into action and, by October 1906, having persuaded the Light Railway Commissioners to delay a decision for a year, they had used the time to draw up draft plans for their own bridge.

In parallel, they conducted a vigorous campaign – leading up to a local referendum – to muster public support against the proposed tramway bridge, claiming that the people of Middlesbrough would have 'sold their birthright' if they did not oppose the plan.

They commissioned the Cleveland Bridge & Engineering Company to draw up plans and specifications for their own bridge – for which the company was paid the princely sum of just £200 – and those plans were used to help promote a Parliamentary Bill, which received Royal Assent in July 1907, clearing the way for work on the present bridge to be started 34 years after Charles Smith had first proposed it.

A detail from one of a series of postcards produced to mark the official opening of the bridge by HRH Prince Arthur of Connaught, 17 October 1911. In the screened-off area in the foreground of the picture, two newsreel cameramen are recording the arrival of the gondola with the official party on board. It would appear that, as well as grandstands at the approach to the bridge, a temporary viewing platform has been created on the first level of one of the towers.

Opposite: A view along the beam towards Port Clarence, with the maintenance cradle in the foreground – used when greasing or replacing carriage wheels.

Invited guests make the first crossing on the Tees transporter bridge in 1911.

Detail of the bridge approach at the Middlesbrough side, from a postcard published in the 1920s.

Progress was reported in the *North Eastern Daily Gazette* on 30 October 1906 who claimed an 'exclusive' insight into the Corporation's plans.

'*The North Eastern Daily Gazette* is in a position to throw an important light on this question by the following announcement:– A definite offer has been made to an important bridge-building firm to construct a bridge of the type decided upon for the sum of £52,000, including the cost of foundations. This sum does not include legal expenses, which would bring the total cost to the ratepayers to about £55,000.

It is extremely interesting to note that the Corporation, in deciding upon a cantilevered transporter, have adopted the principle laid before them so long ago in 1873 by Mr. Charles Smith of Hartlepool.

After exhaustive discussions and inquiry extending over many months, the Middlesbrough Corporation have at last come to a definite conclusion concerning the best means of improving the existing method of communication between the North and South banks of the River Tees. With the aid of expert advice they have carefully weighed the respective merits of transporters – arched, suspension, and cantilever – bascule bridges and improved ferries, and have even discussed the feasibility of a tunnel. They have finally decided on a transporter bridge of the cantilever type, similar to the bridge at Nantes.'

It is somewhat ironic that, having illustrated the article with a picture of Ferdinand Arnodin's cable-stayed Nantes bridge, what was eventually built was much more like Charles Smith's 1873 design. That fact was not lost on the local press – who considered Hartlepool to be part of Tees-side and thus considered Smith to have been a local man whose idea back in 1873 had simply been ahead of its time. *The North Eastern Daily Gazette* delighted in the fact that Smith's ideas had predated those of Palacio and Arnodin by more than 20 years.

Given that it was Georges Camille Imbault who was tasked with designing the new bridge, the reversion to Smith's suggestion for a rigid cantilever rather than Arnodin's cable-stayed alternative was an interesting one. It seems to have been grounded upon concerns about the strength of a large transporter bridge on such an exposed river crossing – exactly the same reasoning which had scuppered earlier plans to build Smith's near-identical bridge.

Imbault, by then Chief Engineer of the Cleveland Bridge & Engineering Company, and a long-time friend and associate of Ferdinand Arnodin, had also overseen the construction of the Newport bridge with its slightly wider span, and would have been well aware of the ability of Arnodin's designs to withstand strong winds. That experience makes it all the more unusual that Imbault's proposed design was nothing like any bridge Arnodin had been involved with.

Cleveland Bridge and Engineering had agreed a contract which ensured that they would be retained as consultants should their tender to build the bridge be unsuccessful. In the event, that is exactly what happened, the construction contract going to Sir William Arrol & Company – whose tender price had undercut them by a considerable amount – with Cleveland Bridge & Engineering receiving a 5 per cent fee to oversee the project.

Arrol eventually agreed a price for the project of £68,026.6s.8d, which, interestingly, is lower than any of the six tenders which the council had originally received. Those had ranged from just over £69,000 to just under £111,000 – and which of them initially came from Cleveland Bridge is unclear.

TRANSPORTER BRIDGE. MIDDLESBROUGH.

Below: Two 6d vehicle tickets hark back to less expensive times – today it costs 60p for a pedestrian to cross on the gondola, or £5 for the full 'visitor experience'.

Below left: A view of the bridge taken from on-board a pleasure craft in the mid-1960s. Moored beneath the bridge is the paddle-steamer tug *John H. Amos*, built in the Paisley yard of Bow, McLachlan & Co. in 1931. Now in a sorry state, the *John H. Amos*, the last paddle steamer tug in British waters, has been raised on to a pontoon at Chatham in preparation for a planned £4M restoration when funding becomes available.

Arrol's was the cheapest, reduced slightly as the design specification for the bridge evolved before construction began.

Needless to say, Cleveland Bridge was not happy about losing the contract, after having done the initial engineering assessments and some of the basic design work. The council as a whole, clearly, was influenced by costs as, even at that late date, their membership was not unanimously behind the project. The almost paternal attitude of some councillors towards the ferries, it seems, had not entirely been abandoned.

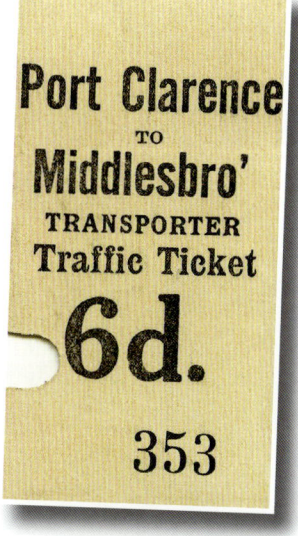

Port Clarence
TO
Middlesbro'
TRANSPORTER
Traffic Ticket
6d.
353

County Borough of Middlesbrough
Transporter Bridge
2192 PORT CLARENCE to MIDDLESBROUGH
TRAFFIC TICKET
6d
Williamson, Printer, Ashton

Despite strong evidence available from the majority of the 12 transporter bridges already in operation across the world, Middlesbrough Council seems to have been somewhat blinkered when it came to realising the full potential offered by the bridge. As a means of transportation, clearly, it offered a significant improvement on the ferries, but any suggestion that the public might find it of interest *per se* was totally dismissed as an irrelevant waste of money. Thus any suggestion that a lift might be installed, and the public charged for the privilege of walking the beam was described by the council's advisors as something which would be 'so rarely used as to be practically valueless and would not justify the expenditure'. A staircase was considered a useful addition – but purely for maintenance purposes.

That contrasts markedly with Arnodin's Marseille bridge with its bistro and gift shop on the overhanging beam, and with Edwardian postcards showing crowds of people promenading across the beams at Marseille, Rouen and Nantes. A lift was finally installed a few years ago.

As might have been expected, so popular was the walk across the top that the 2.5 metre wide steel walkway decking became worryingly worn and had to be covered with teak planking in 1929. Later, that in turn was replaced by a steel grating, but since restoration and the installation of the glass lift giving disabled access to the beam, the walkway is now a durable and wheelchair-friendly fibreglass-based composite.

It remains a moot point whether or not the Arrol quote was ever a realistic reflection of the likely cost of the bridge. The final bill, at £87,300, was £19,000 over budget – a massive 28 per cent. It was £7,300 more than the £80,000 estimated by the Middlesbrough & West Hartlepool Light Railway for their tramway bridge – a figure which had been dismissed as unrealistic by the local council when it had been suggested just a few years earlier.

Arrols retained very tight control over every aspect of the construction of the bridge. The major steel fabrication was all undertaken at the company's Dalmarnock Works in Glasgow and transported down to the site. Given the availability of high quality steel on Teeside, that is just as surprising as French steel being shipped to the iron-rich Basque country for use on the Viscaya Bridge. The Teessiders, however, seem to have made less of a protest over that decision than the Basques had done nearly 20 years earlier.

More than 2,800 tons of steel went into building the bridge, making it by far the heaviest of Britain's five transporters, and while nobody is absolutely certain, it has been roughly estimated that somewhere around 4 million rivets were used in the process – and that is almost two-thirds of the 6.5 million said to have been used on the 2,529 metre long Forth Bridge.

Construction started in July 1909 and took 27 months – and there was a modest incentive for Arrols to complete the project on time – a £50 per week penalty for any overruns. The actual building programme was subject to several modifications imposed by the ferry authority to increase the bridge's carrying capacity which, in turn, required a rethink on the strength of the main beam.

Opposite above left: The original toothed drive from the 500 volt DC motor to the winch. The teeth on the small cog on the motor axle have been cut back since the original motor was decommissioned. The two cables or 'ropes' which drive the gondola can be seen on the cable drum. The long rope measures 518 metres while the short rope measures 280 metres.

Opposite above right: The back-up Westinghouse controller in the winch-house which could be used if the main control system failed. The operator had to keep his foot on the pedal at all times when operating the gondola from this point.

Opposite middle left: One of the original 300amp fuses used in the control panel located at the back of the winch-house.

Opposite below: The winches with the original Westinghouse DC motor, *right*, and the modern 3-phase AC motor installed in 2000, *left*.

Above: A spare wheel for the travelling frame is kept ready on the beam. The solid wheels, which have no bearings, wear relatively quickly and have to be changed quite frequently.

Right: The bridge seen from the approach road on the Middlesbrough shore.

The travelling frame in its 'rest' position, south of the Middlesbrough tower.

That need not have been an issue as, like many Arrol projects – especially the Forth Bridge – the transporter was considerably over-engineered. More than a century later, its complex cantilevered beam remains relatively true – with just a slight kink towards the Port Clarence side and very much less than the distortion of the beam on the Newport Bridge, and it experiences only a tiny deflection as the fully-laden gondola reaches the mid point of its traverse.

We should not be surprised by the bridge's durability, however, given Arrol's reputation – and that his supervising engineer on the project, Andrew Boggart, had also worked on the Forth Bridge.

More than a third of that 27-month project time – and not far short of 40 per cent of the total cost – was taken up with the excavation of the foundations for the tower piers and the building of the concrete base piers on either side of the river.

Rather than topping out the piers above water-level and erecting the towers on top – as had been used for every one of the transporters already completed – the tops of the piers themselves remain below water-level. That is a notable deviation from the three earlier bridges proposed for the site, the designers of all of which envisaged that the foundation piers would be built into the quayside on the Middlesbrough shore, with conventional stone and concrete piers rising above the waterline on the Port Clarence side.

Once work on the erection of the steel towers started, progress was quite rapid, and in just a year the two cantilevered sections of the girder met in the middle and were finally riveted together.

During the assembly of the beam, the increasing weight of steel had been supported in part by two scaffolding towers built under the outer edges of the overhangs and these were dismantled once the two halves were united and the service cabling had been installed.

Thereafter, the stresses have been borne by eight vertical and two diagonal anchor cables from each of the overhanging beam ends, embedded deep in the ground in massive concrete and granite anchorages.

Each of the cables is made up of 127 twisted wire 'ropes' manufactured using a technique pioneered by Ferdinand Arnodin. Deep below ground, tunnels give access to inspection pits so that the condition of the embedded ends of those cables can be regularly monitored.

The travelling frame, from which the gondola is suspended, runs on 60 heavily greased solid steel wheels. These have no bearings on their axles and so experience considerable wear, requiring their regular replacement – how frequently depends, of course, on how heavily the bridge is used and the weights it is carrying.

The gondola – originally designed to carry 600 passengers in the days before heavy motor vehicles were a major consideration – is suspended from the carriage by 30 suspension wires, splayed to ensure that lateral deviation during transit is minimised. Today's gondola is very different from the original installed in 1911 – as are the systems used to operate the bridge.

The gondola approaching its docking station between the Middlesbrough towers.

Opposite above: The view from the gondola deck as it approaches the Middlesbrough docking station.

Opposite middle left: The gondola seen from the main beam.

Opposite below left: The travelling frame wheels as seen from the main beam. For some reason, painting the metalwork of the travelling frame was not included in the specification for the most recent paint job, and it is now in urgent need of attention.

Opposite right: One of the former pulley wheels used to draw the cables – 'ropes' – across the main beam is now displayed outside the Visitor Centre at the Middlesbrough end of the bridge.

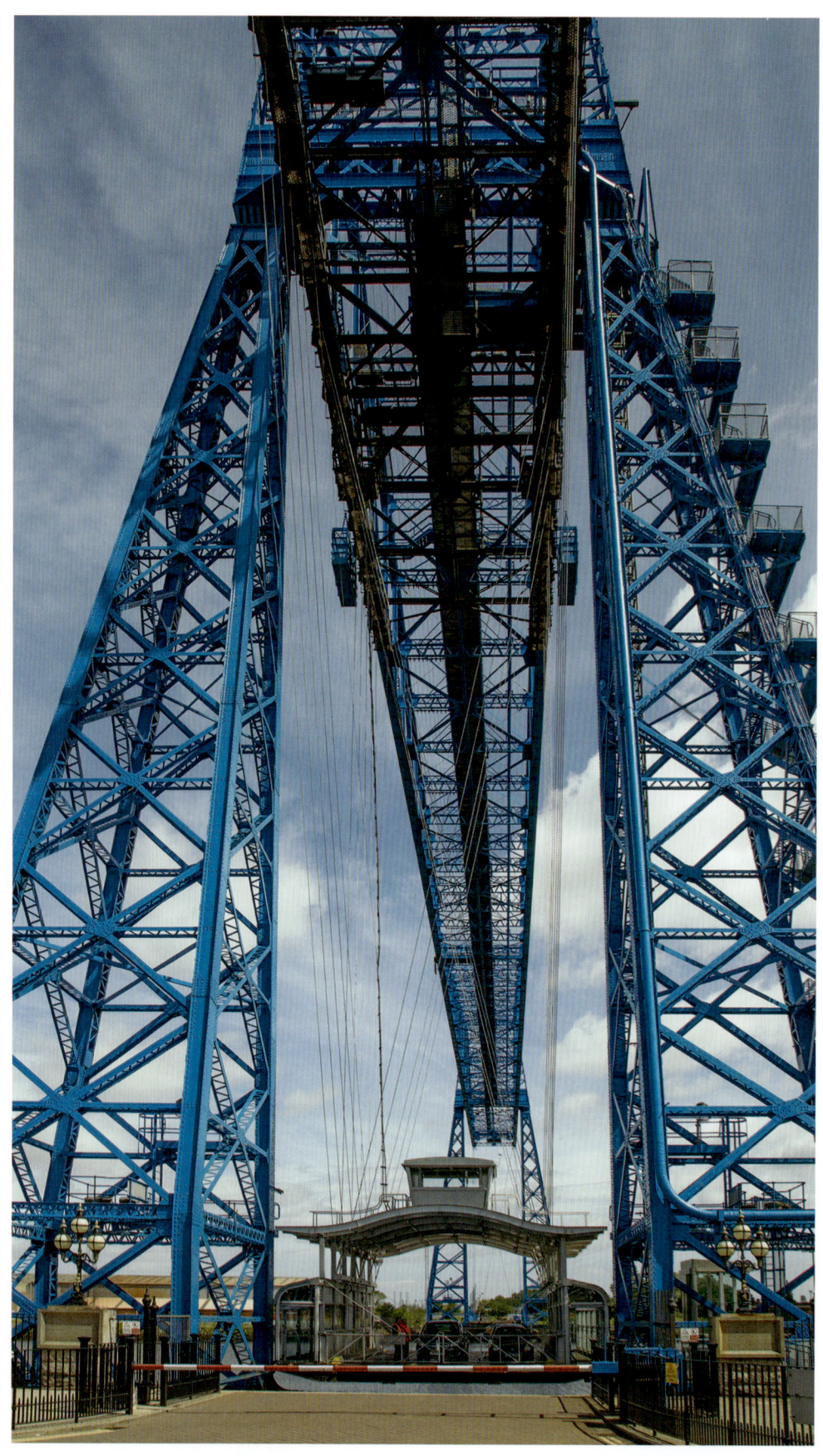

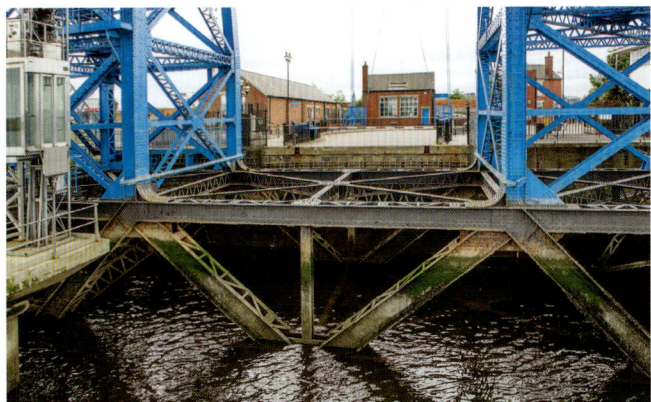

Originally two 30hp Westinghouse DC electric motors provided the power for the winches. Located in the small winch-house on the Middlesbrough side, these motors served for almost 90 years. Since the refurbishment of the bridge in the late 1990s – and in a commendable desire to preserve the bridge's heritage, those motors, now decommissioned, are still *in situ*. Drive now comes from a pair of modern three-phase motors, driving the same winch.

Control of the gondola was originally in the hands of a driver who was housed in a small driving cabin mounted atop the gondola superstructure, using simple tramcar controls also supplied by British Westinghouse from their Trafford Park factory in Manchester.

Electricity for the driving controls, as well as for lighting, was drawn from copper wires which ran beneath the main beam. In the event of a loss of electrical contact, the gondola could be controlled from a back-up controller in the winch-house. But that was then! Today's pilot controls the winch motors using a wi-fi signal from a small handset.

The bridge has had a chequered life, and not without incident. The gondola was hit by a German bomb during the Second World War causing extensive damage, but within two days it was back in service. At least one other bomb – which luckily did not explode – fell into the river and lay there for many years, probably something very few of those who used the bridge daily ever realised.

By the early 1970s, with the bridge already more than 60 years old, a series of cracks were found in the rails from which the overhead carriage ran, and by the middle of the decade expensive repairs were needed which would take the bridge out of service while they were carried out.

Repair costs were estimated at almost twice the bridge's original construction costs, and a vigorous debate ensued about whether or not the cheaper option – demolishing it – made better economic sense. Traffic on the bridge was already waning and, after all, similar financial concerns had led to the demolition of the Widnes–Runcorn bridge more than a decade earlier.

The idea of dismantling the bridge was, eventually, overwhelmingly opposed by the council, but only after a debate about its future – which was

A dredger passing under the bridge. Relatively little shipping uses this stretch of the river today, so ships passing beneath the bridge are now rarely seen.

The narrow maintenance walkway on the overhanging Middlesbrough end of the beam.

Ten anchor cables at each end of the beam support the cantilevered structure.

Above left: The restored Edwardian lamps and commemorative stones which stand either side of the Middlesbrough access ramp.

Right: Since restoration, other uses have been found for the bridge. Bungee-jumping occasionally takes place from the centre of the beam – from a specially constructed platform – and abseiling from the Middlesbrough end of the beam. These activities take place on Sundays when the bridge is not open for cross-river traffic.

every bit as heated as the debates about whether or not to build it in the first place had been in 1908.

The programme of essential repairs proved to be more extensive and more challenging than predicted – every necessary part having to be custom-made.

The timetable was extended by industrial action, and with new rails fitted, the bridge eventually reopened after 15 months.

A further two-year closure followed the shearing off of one of the carriage's running wheels, bringing the total repair costs for the bridge in 20 years to a figure approaching £300,000.

A lot of water has passed beneath the bridge since then, and by the time it became the star of the third series of *Auf Wiedersehen, Pet* in 2002 – albeit looking rather shabby – its future as an iconic symbol of Teeside had been assured. If Charles Smith was to return to Teesside today, even he would instantly recognise the bridge. Middlesbrough without it would be unthinkable.

But such iconic status comes with a hefty and on-going price tag. A large Heritage Lottery Fund grant at the time of the centenary enabled much-needed repairs and improvements to be carried out, and the bridge was closed again for six weeks in 2013 to be repainted in a reflective blue paint – chosen because it looked best under floodlights.

Further investment in the century-old bridge included the long-promised lift. For some reason, however, the repainting schedule did not include the travelling frame, and at the time of writing that is much in need of attention.

It now operates six days per week, and on Sundays, thanks to a specially constructed platform, it is now regularly used for abseiling.

Maintaining the bridge is – and always will be – an ongoing challenge, but however much of an anachronism in the twenty-first century it and the other surviving transporter bridges are, they deserve the additional protection which would come with World Heritage Site status.

Opposite above right: The bridge seen through the ornate wall of the former Bolckow, Vaughan & Company Ironworks on the Middlesbrough shore of the Tees. The company was once celebrated as the biggest manufacturer of pig-iron in the world. Lack of investment eventually led to the company's failure in 1929, its assets being acquired by Dorman, Long & Co. The three major construction companies referenced in this picture – Arrol, Cleveland Bridge and Dorman Long would, in the 1960s, form a consortium to build the Forth Road Bridge.

A considerable amount of work is currently underway to reclaim and restore former industrial land around the bridge, and the local authority has created the 'Transporter Bridge Trail' to encourage more visitors to explore the bridge's historic industrial surroundings.

CROSFIELD'S WARRINGTON BRIDGES

Britain's third transporter bridge was also the country's first to be built on private land and with no public access. It was built to convey materials across the River Mersey to and from the main site of Joseph Crosfield & Sons' chemical and soap works in Warrington to new facilities which they planned to develop on a peninsula site known as 'Tongueland' almost surrounded by a loop in the river.

One of the by-products of their activities was a large quantity of lime, and they sought to make use of this by building a cement works. However, the site they occupied by the Mersey – bounded by the river on one side and housing together with the main west coast railway line on the other – offered little scope for expansion. On the opposite side of the river, however, was that empty tract of boggy land.

When the company started planning this major expansion of their operations, the first step was the acquisition of a 999-year lease on the land from its owner, Sir Gilbert Greenall, later Lord Daresbury, great grandson of the founder of the brewing empire.

Opposite: Photographed by Birtles & Company of Leon House Studios in Warrington, the partially completed tower for Crosfield's first Transporter Bridge on the north bank of the Mersey.

Below: The approximate location of the first bridge is now occupied by this pipe bridge supplying essential services for the chemical works on both sides of the river. The second transporter bridge can be seen in the distance, about 400 metres from the site of the original.

Work underway on constructing the stiffening beam, in the days before health and safety were considerations. In addition to producing industrial photographs of high quality – such as their series on Crosfield's bridge – Birtles operated portrait studios in both Warrington and Northwich.

The partially completed suspension span, photographed by Birtles a few months later from the roof of one of the buildings in the chemical works.

The lease was finalised in 1904 at an annual rent of £158 18s. 8d., and almost immediately, Crosfield's chief engineer, James Newall, was charged with identifying the best way to provide easy access from the existing factory across the Mersey to the proposed site of the cement works.

When work started on the detailed design of the bridge, only five transporters had been completed in the world – with a further three nearing completion at Duluth, Marseille and, of course, the massive Widnes to Runcorn bridge, all three of which would enter service in 1905 just as construction work at Crosfield's was starting.

It would be reasonable to assume that Piggott's engineers would have visited at least one of the bridges which had already been completed – Rouen probably being the most likely – and that they would also have visited the construction site down the river at Widnes/Runcorn.

Right: The bridge under construction as illustrated in the journal *The Engineer* in April 1908, by which time, of course, construction was complete and the bridge was already in service.

350   THE ENGINEER   APRIL 3, 1908

TRANSPORTER BRIDGE AT WARRINGTON

MR. JAMES NEWALL, WARRINGTON, ENGINEER

(For description see page 341.)

Fig. 16.—THE COMPLETE BRIDGE

Fig. 17.—THE WAREHOUSE TOWER   Fig. 18.—PILING OPERATIONS

Below: The partially completed suspension span, also from the article in *The Engineer*. Comparing this published version of Birtles' photograph with the original – *see earlier* – the two men hanging precariously on a single wire beneath the main suspension cable just to the left of the chimney have been edited out. Removing the men seems an odd decision – especially in the days before health and safety considerations carried much weight – as there would have been no other way of fixing the vertical cables, despite the obvious and considerable risks involved.

Ideas for their bridge were clearly influenced both by Arnodin's Rouen design and the technical innovations of John Webster and John Wood's Widnes–Runcorn crossing, which was under construction by the Arrol Brothers through their Arrol Bridge & Roof Company of Germiston Ironworks in Springburn, Glasgow.

As the projected performance requirements for the planned bridge were somewhat limited in terms of carrying capacity, a relatively simple lightweight suspension structure with a single stiffening beam would be chosen.

All the world's other transporter bridges would be built using a design which consisted of two widely-spaced stiffening beams cross-braced with a lattice of girders as an aid to rigidity.

The reduced weight of the stiffening beam, in turn, meant that the towers could also be of a very simple and lightweight winged pylon design.

Despite being the smallest transporter bridge yet constructed, with a span of just 76 metres when completed in late 1907, the bridge attracted a great deal of interest from engineering journals.

The Engineer in its issues of 25 March and 3 April 1908 gave a detailed account of its design and construction as well as an historical perspective on those few transporter bridges which had, by then, already been constructed around the world.

'Several schemes for establishing communications across the river were considered; any form of ferry would be most inconvenient on account of the rise and fall of the tide, and the extremely rapid tidal currents; a tunnel

The original transporter stood for many years after the completion of the second bridge, but records do not tell us whether or not it was used regularly after that time. It is seen here at low tide in 1960, and it was certainly still standing in 1962. Its longevity may have had nothing to do with its original function, but rather because of the more fundamental fact that it carried pipework and cabling across the river on the stiffening beam – functions now fulfilled by the present-day pipe bridge which crosses the river at the same point.

The south tower was built on the roof of one of the factory buildings in the original soap and chemical works. Assuming the height of the figures to be around 1.7 metres – 5ft 7ins would have been a typical height a century ago – the tower can be estimated at around 17 metres. The two 'wings' were attached to stabilising cables anchored in the yard before the bridge's main suspension cables were slung across the river.

was impractical, for no approach could be arranged in the old works, and the idea of entering a tunnel by shafts with hydraulic and other lifts was considered unsuitable on account of the time required and working expense. The only alternative left was to adopt some form of bridge which, whilst meeting the requirements of the works, would not interfere with the navigation of the river.'

To build the bridge, Newall enlisted the services of Thomas Piggott & Company Ltd. of Birmingham, whose design was considered the most suitable of those submitted as part of an open competition. As has been mentioned, Piggott had come up with a highly practical but much lighter structure than ever built before.

To record the construction process, Piggotts enlisted one of Warrington's longest-established photographic studios – Birtles & Company of Leon House – which had been established by Thomas Birtles both in Warrington and in nearby Northwich as early as 1864. Birtles, one of the first students to attend Warrington's School of Art – founded in 1857 – had, by the time the Crosfield photographs were taken, already been awarded the prestigious Fellowship of the Royal Photographic Society. Several of the photographs commissioned by Piggotts survive in the company's archives, now held in Staffordshire Country Record Office.

From those photographs it is clear just how simple the structure was – especially when compared with what was required to support the Rouen bridge's 143-metre span, and the Widnes–Runcorn bridge's 304 metres.

Piggott's design for the bridge envisaged a maximum load on the gondola of only 2.5 tons, and a speed of 6mph – a crossing time of just 30 seconds – and that meant that the stiffening beam did not need to be anything like as robust or heavy as would have been needed for a wider span. The design, therefore, required only single towers on either side of the river whereas every other bridge built thus far used double towers. Such a design, however, imposed severe restrictions on the bulk of the objects being transported as well as their weight.

The layout of the existing works also imposed its own restrictions, an issue neatly circumvented by building one of the towers on top of existing warehouse buildings, materials being loaded on and off the gondola directly in the works yard. On the new site, a wooden jetty constructed by the river bank served as a loading and unloading bay.

In designing the bridge – and especially its stiffening beam – Piggotts drew on the widely published work of eminent bridge engineers Mansfield Merriman and Henry Sylvester Jacoby whose *Text-Book on Roofs and Bridges* was considered to be the standard reference work when evaluating and calculating the stresses to which the structure would be subjected under load.

The Engineer magazine noted that for the Warrington bridge, Piggott's design engineered considerable strength into a remarkably simple and lightweight structure. The design considered the most effective and economical cable

Right: An evocative view of the quayside showing the old works with steam barges loading or unloading their cargoes, and the transporter bridge's main beam soaring overhead. This view was one of a series of professional photographs of the chemical works taken in May 1920. (Reproduced with kind permission of Unilever from an original in Unilever Archives)

Below: Also from the May 1920 series, this photograph gives a tantalising glimpse of one of the locomotives working the Tongueland site's railway. The 0-4-0 saddle-tank locomotive is of a type built by Hudswell Clarke of Hunslet, Leeds, to a late nineteenth century design. There is no record in the Hudswell Clarke sales books of it having been supplied new to Crosfields, so it may have been bought secondhand, specifically because its weight – just 15.5 tons – allowed it to be transported within the weight limit of the original gondola on the 1916 bridge. The company continued to manufacture these little locomotives at least until the early years of the twentieth century. (Reproduced with kind permission of Unilever from an original in Unilever Archives)

dip – the vertical difference between the height of the tower and the centre of the span – to be one twelfth of the 250ft span or around 20 feet, and:

'…as it was decided to make the cables continuous from anchorage to anchorage, the maximum stress required a cable of 7in circumference. These cables were made of the best selected plough steel wire, each wire having an

ultimate strength equal to 120 tons to the square inch. As there were in each cable seven strands, each containing thirty-seven wires, its total strength came out at just over 200 tons, the guaranteed strength being 180 tons.'

As the design used single towers on either shore, the main suspension cables were uncradled. This would have meant that without additional stabilisation, the suspended stiffening girder – itself of relatively lightweight construction – would have been prone to both lateral and vertical deflection. With a cradled design, as the distance between the cables narrowed towards the centre of the span, lateral flexing was markedly reduced.

A wider twin-tower structure would have facilitated cradling and thus obviated that problem, but would, of course, have had to be built to a much more robust and expensive design, but the designer minimised lateral movement at Crosfield's by fitting 'wings' to the towers which were stabilised by radial steel cable stays buried deep in the yard on both sides of the river.

Back stay cables behind each tower added to the strength. Flexibility must have caused problems both with the bridge carriage's unusual direct drive system, and with the beam's dual role as a conduit for cables and pipework between the two sites. It was also fitted with a wooden walkway for maintenance.

Construction work is believed to have started in the spring of 1905 and to have lasted around 18 months – including the piling work which was necessary to create the concrete caisson and quay on which the tower on the

The type of Hudswell Clarke locomotive seen in action in the photograph of Crosfield's site. This design, first introduced in the 1880s, was kept in production for many years. This class of locomotive was widely used by contractors on construction sites and could negotiate curves with a radius of just 35 feet. One example of the class survives in the UK. Known as *Lord Mayor* and built in 1893 with Works No.402, it has been preserved by the Vintage Carriages Trust at the Keighley & Worth Valley Railway in Yorkshire.

Right and below: OS Maps showing the peninsula before and after Crosfield developed the site and its railway network. The maps are dated 1905 and 1926. Both transporter bridges are shown on the 1926 map, although only the 1916 bridge is identified. The 1907 bridge can be seen crossing the Mersey at the top of the map.

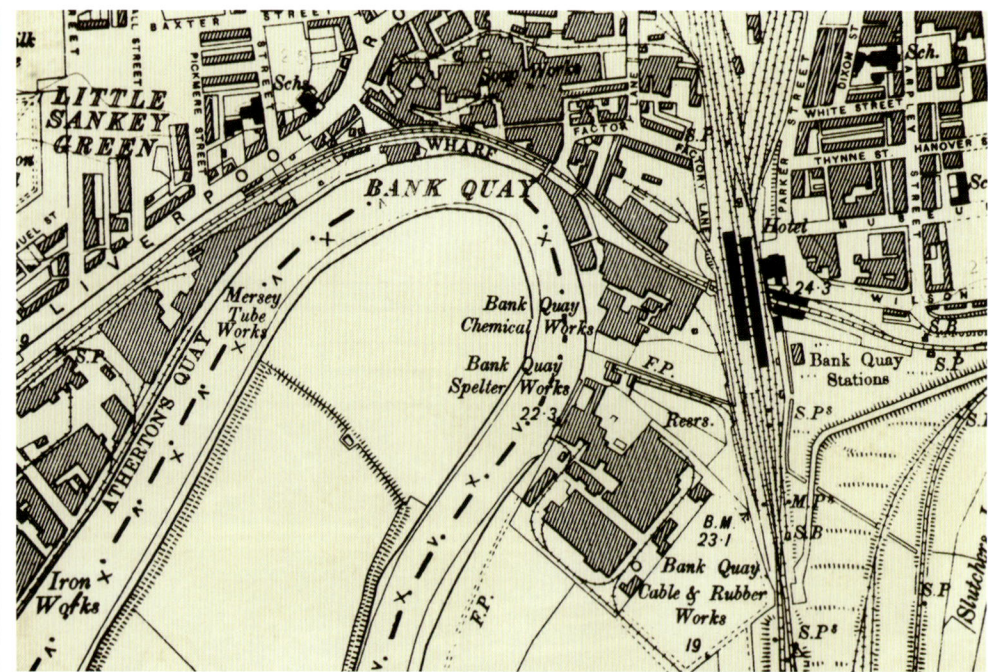

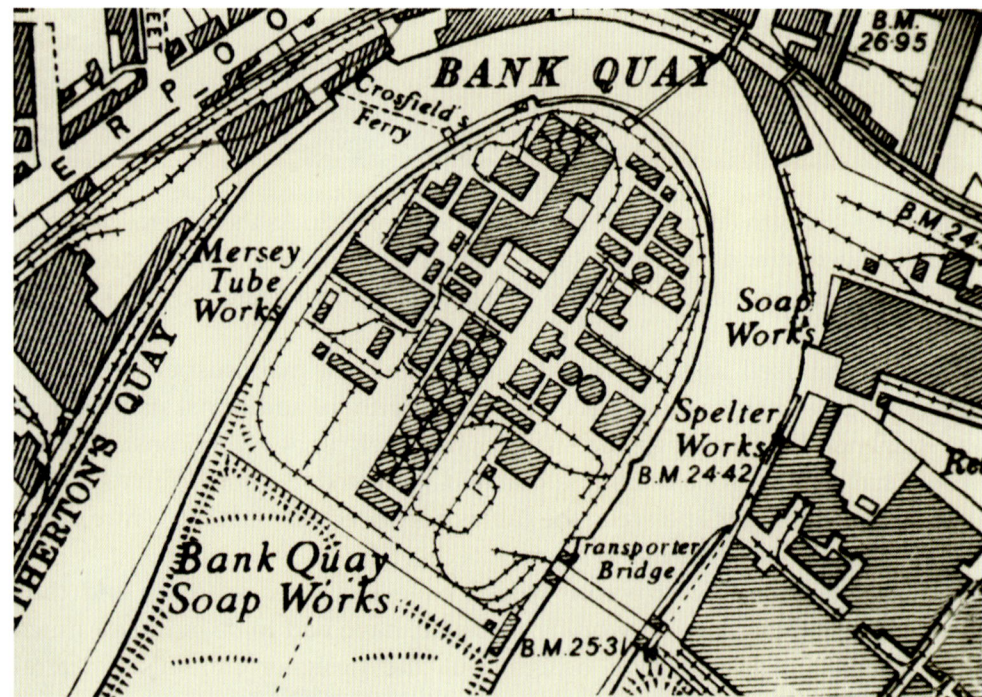

Cheshire side of the river was built – with the bridge entering service towards the end of 1907.

In the end, the total construction cost was reported to have been just £4,000 – a tiny figure when compared with the sums which would have been required to fund either a high level bridge or a tunnel, and the equally massive sums which had been required to built any other of the world's transporters.

Perhaps impressed by the revolutionary drive system which had been installed on the Widnes–Runcorn bridge, shortly before construction work on the Crosfield bridge started the decision was made to eschew the more conventional cable and winch system for the travelling frame, and use DC electric motors to directly drive the bogies.

On the Widnes–Runcorn bridge, despite its massive strength and rigidity, engineers would report power loss as the flexing beam caused a break between the live rail and the brushes which picked up that power and delivered it to the motors. In 1913, just six years after the Crosfield bridge entered service, Widnes Corporation abandoned that system and resorted to winches and cable drive for their bridge. The designers of the Crosfield bridge, however, believed that their much simpler design would eliminate such problems, and there is no surviving evidence to suggest they regretted their decision or felt the need to modify the drive system. *The Engineer* report continued:

> 'In the Warrington bridge a continuous current of 220 volts is conveyed to the trolley by two wires arranged on either side of the runway, the current being picked up by means of a couple of small trolley arms similar to those used in street tramway work. A tractive force of 100lb. per ton was allowed and the trolley … has its four wheels coupled by means of sprockets on their axles and a chain drive connection.
>
> It is well to avoid a multiplicity of wheels in the trolley on account of the uneven bearing they would take through deflection of the girder.'

The result was very effective in all respects except its limited carrying capacity, and the report in *The Engineer* ended with a testament to Piggott's design and the bridge's rigidity.

> 'After completion, the bridge was tested for deflection under working load, and this came out extremely small, being only 1½in. at quarter span, and 2¼in. in the centre.'

Opinions differ as to how long it remained in use or when it was dismantled. Some sources have claimed that it was rarely used after the 1916 bridge opened, but available records offer no clarity on that point. If that is the case, however, questions arise as to why it was not dismantled for at least another 45 years. It survived at least until 1962, its site eventually being taken by the present-day pipe bridge.

Not long after the bridge's completion, Crosfields bought the land they had leased from the Greenalls, paying a reported £3,726 for it, and developing a number of different chemical facilities on the site. To facilitate transportation of raw materials and finished products around the site, they also planned to build a standard-gauge private railway which would link several of their different facilities.

While the railway system would offer access to the riverside wharf on the peninsula or 'tongue' site, connecting it to their established premises across the river – and thus to the national rail network – was one of the major considerations behind the plans to construct a larger and stronger transporter bridge about 400 metres further along the river.

In 1909, just 18 months after the first transporter bridge was completed, the company acquired the UK rights to manufacture Persil, the revolutionary new German washing detergent. It would go on to become their – and subsequent owner Unilever's – best-selling product. Production on the Warrington site was destined to rise considerably.

By 1911, however, Crosfield's days as an independent company had come to an end, the entire business having been bought by the Northwich-based chemical conglomerate Brunner Mond & Company – a company with which John Crosfield had a financial interest and a long association, having served as their chairman for several years.

It was under their ownership that the decision was made to construct the new bridge. Plans were drawn up around 1912, by which time there was a much greater weight of raw materials to be moved between the sites.

To design the new bridge, the company – still trading as Joseph Crosfield & Sons but recently acquired by Brunner Mond – commissioned the widely

A tanker lorry on the gondola in the 1940s, before it was extended. (Reproduced with kind permission of Unilever from an original in Unilever Archives)

respected William Henry Hunter, Consulting Engineer to the Manchester Ship Canal Company.

Hunter was also Commissioner for the Upper Mersey Navigation and in that role was only too well aware of the challenges of navigating even the upper tidal reaches of the Mersey.

In his book *Rivers and Estuaries or Streams and Tides* published in 1912 just before he took on the Warrington design brief, he had noted that even as far up stream as Warrington, the river could rise by as much as 1.2 metres at spring high tides. The strong currents as the tide ebbed and flowed were a well-known challenge to the generations of ferrymen who had worked the river for centuries.

Hunter came up with a cantilevered bridge of immense rigidity and strength – some might say over-engineered, but then that was a characteristic of many of contemporary engineering projects. That robustness would later make it possible to increase the bridge's capacity.

Perhaps aware of their work on the iconic Forth Bridge – also said to have been heavily over-engineered – as well as the internal structural steelwork for London's Tower Bridge, and the Tees Transporter Bridge in Middlesbrough which had been officially opened just the previous year – and on which they had worked with the Cleveland Bridge & Engineering Company as co-erectors – Crosfields gave the contract for their new bridge to Sir William Arrol & Company of Dalmarnock Iron Works in Glasgow.

A 'Perfection' railway wagon mid-stream in 1920. *Perfection Soap* tablets had been introduced by Crosfield in the 1880s. (Reproduced with kind permission of Unilever from an original in Unilever Archives)

The transporter bridge was placed on the 'Buildings at Risk Register' by Historic England in 2003. Crosfield planned to demolish it in 1972 but were thwarted by its listing as an Ancient Monument in 1976. Forty years later, the present owners of the site severely restrict access to the bridge despite its listed status. The cost of carrying out essential repairs and painting the bridge to ensure its future survival is now said to be approaching half a million pounds. Compared with the photographs on the previous pages, it can be seen that over the years the bridge piers, once sitting at the edge of the river bank, have been almost overwhelmed by undergrowth and are now well back from the much narrowed channel.

At just 61 metres, the span Arrols were contracted to build was significantly less than Piggott's 1907 bridge, but the weight it would be carrying demanded a much more robust structure.

The brief required that the bridge have a carrying capacity of 18 tons and it was to be fitted with rails to allow the passage of railway wagons and small locomotives between the two sites. Extension of the railway – which already served the main works – across the bridge and around the peninsula site was a key part of the company's expansion plan.

It is therefore highly unlikely that the rail network was in existence before the second bridge was completed as there was no practical way in which

The underside of the travelling frame looking towards the north end of the bridge, just below the winch-house.

The return point for the cable drive at the south end of the cantilevered stiffening beam.

locomotives and rolling stock which would operate on it could have been delivered to the site in the years when the weight limit on the earlier bridge was just two and a half tons. Looking at contemporary maps, it is apparent that the railway was built with some very tight curves.

Crosfields were already using steam locomotives in their works long before even the first bridge was built. One 0-4-0ST locomotive – carrying Works No.658 and named *Perfection* after their popular soap – had been ordered new

The 1948-built 150hp Class 420 locomotive *Geoffrey Heyworth* – named after Unilever's Chairman – was the most powerful locomotive in the fleet.

The last Fowler shunter to work on the site was *Persil*, a 1952-built 100hp Class 416 0-4-0, named after the famous detergent. Moved to Southport's Steamport in 1971, *Persil* is now displayed as a static exhibit in Preston's Ribble Steam Museum.

A rail wagon on the extended gondola, photographed as it approached its docking bay between the bridge towers on the south side of the river. The metal platform for road vehicles can be seen beneath the wagon. While the width of 7.3 metres was an obvious necessity, it is hard to understand why the internal height clearance needed to be so great. (Reproduced with kind permission of Unilever from an original in Unilever Archives)

from Hudswell Clarke of Leeds on 20 March 1903 and delivered to the site on 29 May. A second identical engine – carrying Works No.979 and named *Crosfield* – was ordered on 15 May 1913 and delivered just four weeks later – months before plans for the second bridge were even off the drawing board. *Perfection* cost £1,145 in 1903, while ten years later, *Crosfield* cost £1,310. Both of these 22-ton locomotives were supplied in the 'Blackberry Black' livery of the London & North Western Railway, suggesting that they would have had running rights over that company's mainline system.

These little locomotives, because of their short wheelbase, could negotiate curves with a radius as tight as 45 feet – and some of the track on the new site had curves with which longer wheelbase engines could certainly not have coped.

The railway layout may have been designed with the capability of these locomotives in mind and, once laid, the curves would have been one of the factors which dictated the specifications of all future steam or diesel engines purchased by the company.

While it remains unclear just how many other locomotives Crosfields owned at the time the second bridge was completed. At least one, a third and older Hudswell Clarke 0-4-0ST (visible at one side of the illustration on page 194), was moved over to work on the new site.

Given that the gondola initially had the capacity to transport just a single railway wagon at a time – which would be pushed on by one locomotive, and pulled off by a second on the other side – there was no requirement that the locomotives on the new site be very powerful. The Hudswell Clarke locomotives, with a weight of just 15.5 tons, were powerful enough for the job, and light enough to use the bridge.

It would appear that steam traction was already being phased out, at least in part, before the start of the Second World War, to be replaced by diesel. That poses further questions. Of the eight diesels which the company – a subsidiary of Unilever since 1919 – subsequently bought from John Fowler of Leeds between 1939 and 1952, only the first came in under the bridge's maximum load capacity. That locomotive – with Fowler's Works No.21909 – was a 60hp shunter named *Charles Huffam* weighing in at just 15 tons.

It arrived on site on 31 January 1939. At some later date, when the locomotive originally named *Mersey* was withdrawn, *Charles Huffam* was renamed *Mersey*.

The whole nature of the transportation of bulk raw materials was changing even before the outbreak of the Second World War, and the decision was made sometime between 1938 and 1940 to modify the gondola platform by installing metal running plates around and between the railway lines, thus permitting the transit of road vehicles as well as railway wagons.

From records in the Unilever Archives, it is evident that those modifications were little more than a stop-gap measure, quickly found to be inadequate, and shortly after the end of the war, further changes were clearly necessary.

With the rapid growth in the size of lorries, especially chemical tankers, used by the company and its suppliers, the gondola was extended some time around 1949. That enlargement of the deck area increased its length by approximately 22 per cent, and it seems probable that it was at that same time that the carrying capacity was increased from 18 to 30 tons. The caption on the photograph, dated 1949 and part of Unilever Archives, records that:

> 'Movement records over recent years have shown that of vehicles using the bridge, the number of road tankers now greatly exceeds that of rail wagons, and the platform has been enlarged to accommodate the ever increasing length of the former type.'

By 1949 six new diesel locomotives had been delivered, all but one of which exceeded the original 18-ton limit – the 20-ton *Metso* in August 1939, *Mersey* (20 tons) in February 1940, an un-named 20-ton locomotive in 1947, and two 29-ton Class 420 150hp diesels – one in 1947 and *Geoffrey Heyworth* in 1948. Assuming that the weight increase had been part the same project which extended the gondola, all the company's locomotives would have had free access across both sites. Another 20-ton locomotive arrived in 1950 and the last, the 100hp 19-ton *Persil* which was delivered in September 1952, was also well under the increased weight limit.

Use of the transporter dwindled rapidly after a new road bridge was opened offering vehicle access across the river and, as a result, it has not been in operation since around 1964.

Arguments about its future have continued ever since, and its condition has become an ever greater cause for concern over the past half century and more.

A little over a decade after it was last used, and with its condition already causing some concern, Warrington Development Corporation offered a grant of £2,700 towards restoring the bridge as a working tourist attraction. A year earlier Crosfields had rung alarm bells by making it clear that unless some public body took over the maintenance of the bridge, it would likely have to be dismantled.

Local newspaper reports told how Cheshire County Council had 'assumed responsibility for the restoration and management of the structure in perpetuity' so hopes were high.

Restoration costs were estimated at £30,000 – almost as much as the £34,000 the bridge had cost to construct more than half a century earlier – and the Department of the Environment agreed in November 1976 – shortly after the bridge had been scheduled as a Grade II* listed monument – to immediately contribute £3,500 to make the structure safe. The Job Creation Programme operated by the Manpower Services Commission had likewise committed to making a substantial contribution towards carrying out the work. But little immediately happened, and by the end of the following year progress seemed to be at a standstill despite everyone agreeing that the bridge was of considerable historical importance and had the potential to be an important tourist attraction for the town.

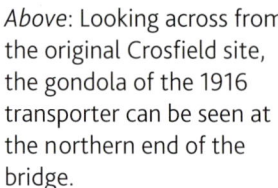

Above: Looking across from the original Crosfield site, the gondola of the 1916 transporter can be seen at the northern end of the bridge.

Above right: The winch-house sits at the northern end of the stiffening beam, overhanging the newer half of the chemical works.

Right: The travelling frame mechanism slung below the beam.

Then, on 28 March 1979, the *Warrington Guardian* ran a feature headlined 'Transporter Bridge will be open for public trips' which seemed to promise an assured future for the bridge.

> 'One of Warrington's best-known landmarks, the Crosfield's Transporter Bridge, which spans the Mersey at Bank Quay, should grind into motion this summer after standing motionless for about 15 years. And members of the public will be able to take rides on the transporter platform, which used to carry railway wagons across the river between the old Lancashire and Cheshire works of Joseph Crosfield & Sons.'

There were notes of both optimism and caution in the report. The Factories Inspectorate had declared the structure safe for public access, and it was claimed that the new paint job which the bridge had been given would 'give it protection for around 110 years'.

The cautions were to do with the challenges of restoring an electricity supply to the bridge – the supply had been discontinued after it was closed in 1964 – and the likely need to commission specialist manufacturers to make replacement parts for the winches and their electric motors.

Cheshire County Council's Conservation Officer, Oliver Bott, quoted in late 1979, was confident that 'This will represent the final phase of the work, and

The gondola is now in such a dire state that restoration would represent a considerable challenge.

we are hopeful that it will be completed in time for members of the public to ride in the transporter this summer.'

That level of optimism was probably more to do with ambition than any realistic expectation, as there had been only limited consideration of the long-term challenges of operating a public tourist attraction safely in the middle of a potentially hazardous chemical works.

Seven years later, *The Guardian* published a photograph of the bridge, giving brief historical notes and captioned:

'But the bridge – now a listed building – remains a reminder of the days of blood, toil, tears and sweat that the labour-saving world of today has evolved out of existence. Soon, someone may turn it in to another tourist attraction, a somewhat undignified end for a veteran workhorse but infinitely better than a scrapyard fate.'

Almost 40 years after its preservation was first talked about, it is clear that confidence in such a plan was sadly misplaced. But the struggle to try and ensure the bridge's future goes on, valiantly championed since March 2015 by

Above and opposite: The effect of the weather is apparent from these two views of the north tower. From the east, the bridge's condition seems pretty good, but well below what might have been expected given the 1979 assurance that the metalwork had been given 'protection for around 110 years'. From the west, the condition of the structure is a worryingly different story.

Too late to save? Let's hope not. Arrol's great engineering achievements are protected and celebrated throughout the country – this one deserves no different.

the admirable 'Friends of Warrington Transporter Bridge' whose mission is to become a strong independent voice both to promote and publicise the bridge, and do whatever can be done to help towards assuring its preservation.

Along with the Keil bridge in Germany it was one of only two transporter bridges designed to incorporate railway lines. The Duluth bridge had been designed to carry passenger tramcars, so Crosfield's bridge, built specifically for railway access is a truly unique survival.

The Historic England listing for the bridge cites it as only of 'Category C' priority, despite accepting that its condition is 'poor' and experiencing 'slow decay' with 'no solution agreed'.

As the condition of the bridge today continues to cause concern, it has been added to the 'buildings at risk register' – a truly sad reflection on all the official plans and aspirations of creating a visitor centre and offering public access to it. As metal continuously deteriorates, time may be running out for this important structure – the only transporter bridge ever built for purely industrial traffic.

In 2016, the bridge was shortlisted for a Heritage Award and presented with a Special Award for Civil Engineering Heritage by ICE North West. Through

The modified and gated gondola today. Its current condition is a shocking indictment of all those public bodies charged with the protection of historic monuments who pledged themselves to its restoration forty years ago – and spent a substantial amount of money in preparation for its public opening – and all those who had stood by as its condition has deteriorated, now perilously close to being beyond rescue.

a close relationship with Warrington Borough Council, who hold the lease on the bridge, brown tourist signs have been put up on the A5061 and Slutchers Lane together with way-marked posts showing the bridge in silhouette. Last year the Friends persuaded the council to erect an information board in Bank Park some distance from the site telling the bridge's story, as well as a finger-post vaguely pointing towards it. The Council Footpaths Officer is keen to improve things, but access remains difficult with parts of the route not for the faint-hearted.

Surely such an important piece of history demands better treatment from a country which profits so much from its industrial heritage

REBUILDING ARNODIN'S ROCHEFORT BRIDGE

Of Arnodin's five transporter bridges to be completed in France, only one survives – the bridge over the Charente at Rochefort-sur-Mer – and its restoration, which involved replacing everything except the towers and the gondola, has offered a unique opportunity, in effect, to see a transporter bridge being built.

When it reopens in 2020, 120 years after it originally entered service, the gondola will once again carry traffic – albeit just pedestrians and cyclists in deference to the age of its towers.

Today's motor traffic is carried by a 1991 high-level road bridge, itself a replacement for a 1966 vertical lift bridge.

Work started on building the transporter bridge in 1898, with the building of stone and concrete piers to what would become Arnodin's stock design – examples of them can still also be seen at Bordeaux, Nantes and, of course, Newport.

Off-site, construction of the towers was undertaken at Arnodin's foundry in Chateauneuf-sur-Loire. They were then dismantled and then reassembled on the banks of the Charente. That reassembly took just four months.

Once the main cables had been slung over the towers and tethered in their stone and concrete anchors set back from the towers, construction of the main beam itself was completed in seven months and the bridge was finally opened to traffic on 29 July 1900. What stands today, however, hardly reflects the open elegance of Arnodin's design.

As with Arnodin and Palacio's Bilbao bridge, only the towers at Rochefort are original – the beam was replaced in the 1930s when the metalwork began to develop stress fractures in the side-frame. At the time, it was concluded that the inherent weakness in the design was at the junction between the cable-stayed section, and the central suspended section of the beam where the stresses were greatest as the gondola passed below.

Whether or not similar stresses would have manifested themselves in the Puente Bizkaia at Portugalete will never be known as the original beam was destroyed during the civil war.

The engineers working on the restoration project at Rochefort today believe that those weaknesses were inherent in the structural design Arnodin used for the bridge, and that even with the much greater strength of modern materials, there remained the remote possibility that the replacement beam might one day develop similar faults. How they resolved that issue is explored later in this chapter.

Above: The Rochefort bridge was Arnodin's first solo transporter bridge project. His bust looks out from near the Echillais Visitor Centre towards the bridge.

Opposite: The Rochefort tower of the Rochefort–Martrou bridge, photographed from the Echillais side of the Charente, the scaffolding carrying the lift shaft already in place in May 2017 as work got underway on the €30M reconstruction of Arnodin's last surviving *pont transbordeur* in France. The 1933 suspension cables were then still in place supporting the beam. The bridge is currently scheduled to re-open in late 2019 after more than three years' work.

A view of the bridge from the Echillais side of the river.

Right and below: Two postcards of the Rochefort–Martrou bridge c.1910. The rebuilding will restore Arnodin's original profile for the bridge's main deck beam.

Rochefort-sur-Mer — Le Transbordeur de Martrou

27 ROCHEFORT. — *Pont de Martron.* — LL.

The 1933 solution to the problem was to replace Arnodin's elegant 2-metre high lattice beam with a heavier plated 1.4-metre high beam with solid sides. It was also in 1933 that the cable stays were replaced with vertical suspension cables – another modification designed to even out the stresses.

Additional anchor cables were put in place, tethered to newly installed 250-ton concrete blocks behind the 1900 anchors.

As a 'Monument Historique', one of the primary requirements of the current project was to restore the bridge as close as possible to Arnodin's original design, thus these additional cables and anchors were removed during the restoration work.

In 1900 the completed bridge had used over 700 tons of iron and steel, including the 175.5 metre over-shooting main beam.

Between the pylons, the span was 140 metres at the beam, reducing to 129 metres at the base piers.

While there was always a walkway across the beam, access to it was originally intended to be just for maintenance purposes, so the staircases within the towers are really more like stepladders. This was still the period in his thinking when Arnodin could not contemplate why anyone would want to climb the towers and walk 'over the top'. Thus there is no substantial staircase within the towers and that, in a way, is part of why the towers seem much lighter in construction than those, at Newport, his only other surviving transporter bridge.

The plated beam from 1933 will be replaced by a recreation of Arnodin's original beam with its open lattice design. The new lattice beam will be 2 metres tall – as was the original – compared with the 1.4-metre plated beam.

Right: Temporary support cables being installed on the Rochefort side of the river in preparation for the removal of the 1933 beam.

Below: The cable anchorages on both sides of the river are being refurbished and will eventually house the embedded ends of the new cables.

From the outset, traffic across the bridge was considerable, with 115,000 vehicles, 675,000 cyclists and 260,000 foot passengers crossing in the first year. A decade later the annual load had more than doubled, possibly exceeding both Arnodin's predictions and the stresses with which the bridge had been designed to cope – more than 106,000 cars, 10,600 lorries and half a million foot passengers. Between 1930 and 1960, the number of cars using the bridge annually grew to more than 400,000.

Long before then, however, the bridge started to develop potentially severe structural issues. Arnodin had evolved a unique design for the main beam to increase its rigidity – a design he also employed on some of his other bridges.

The 2-metre high sides of the beam were stiffened with bolted rigid steel cross-braces, and within 30 years of its opening, these braces were exhibiting stress fractures. The decision was made in 1933 to replace Arnodin's elegant (if flawed) design with the new plated beam.

At the same time – and perhaps because of the substantially increased weight of the new beam – the suspension system was radically redesigned.

In Arnodin's original design, the majority of the stresses on the structure were directed downwards through the towers, the beam largely being

Seen from the top of the Rochefort tower in May 2017, the workers on the suspended platform are dismantling the plated sides of the beam, watched by a colleague in protective clothing who has been removing asbestos paint from the beam itself.

Above left: A 1907 postcard of the bridge, looking towards the Rochefort tower and winch-house. The bridge is cable-driven from the Rochefort side. The winch-house, which originally housed a steam engine is at the foot of the right-hand tower. The current electric motors date from 1994, replacing those installed in 1927.

Above right: The Rochefort tower in May 2017, viewed along the same roadway – then being used as the construction yard.

supported by cable-stays radiating from the top of each tower. Only the central section was supported by suspension cables. The 1933 modification replaced the cable-stayed design with suspension cables across the full span and the effect of that was to divert some of the stresses from the towers and transfer them to the anchor cables.

Dealing with that required additional cables to be slung between the towers and the anchorages, embedded in those new blocks of additional concrete behind each of the anchorages. Much of the three-year programme has been taken up with undoing those alterations.

As part of the restoration, those additional anchor cables have been removed – along with their concrete anchorages – and replaced with the lighter-weight beam and cable-stayed suspension system that Arnodin designed back in the 1890s.

The gondola – a replica as the original had been destroyed by the Germans in 1944 – made its last crossing on 4 February 1967, less than a year after the bridge had been a feature of Jacques Demy's film *Les Demoiselles de Rochefort* – and it seemed that demolition would soon follow. For years it stood idle,

Arnodin's name is cast into the metalwork at the very top of each tower – where virtually nobody ever sees it.

Left: The vertical suspension cables were introduced in 1933, slung from anchorages bolted across the four main cables on each side of the beam.

Below left: Arnodin and his bridge.

Below: This crude-looking device dates from the early years of the bridge. Known as the 'bicyclette', it was suspended from one of the main cables while a worker, suspended from it, periodically applied a coating of tar to protect the cables from corrosion.

Pont transbordeur de Rochefort - Ferdinand ARNODIN

gradually deteriorating. A sum of 1.4 million francs had been set aside to meet the cost of its demolition, but thankfully that plan was never put into effect. By that time it was the last example of Arnodin's transporter bridges still standing in France, and its recognition and preservation as an historic monument should have been assured.

Eventually, in 1976, it was afforded that status, and the funding put aside for its demolition was instead applied to halting its deterioration.

Work on consolidating the bridge started in 1981 with a major refurbishment initiated in 1990. In 1994 the bridge re-opened as an historic tourist attraction, but now carrying only cyclists and pedestrians. A visitor centre opened on the Echillais side in 2003, since when the bridge has attracted half a million visitors. It has remained open during the current restoration.

By 2010, however, there were growing concerns about the bridge's integrity. Extensive corrosion resulting from inadequate maintenance and sporadic painting had left the structure measurably weakened, and a rupture in one of the anchor cables in 2010 seriously threatened its future.

That led to a detailed structural survey of the entire bridge which identified the scale of the challenge which engineers would have to tackle. It was a bold decision to even consider a major restoration of the structure and, in these straitened times, quite remarkable that the necessary funding for such a colossal project was made available relatively quickly.

Extensive research was undertaken on the logistical challenges which such a project presented, and that was followed by computer modelling of possible

Opposite: May 2017 – a dramatic view, looking towards Echillais, from the temporary crane platform 67 metres above the Charente which was erected on top of the north tower of the bridge. The temporary walkways facilitated the installation of the temporary support cables and, later, the new main cables and the fixings of the new suspension cables and stays. The access elevators were put in place early in 2017 and at the time this photograph was taken, work was already underway to remove the plated beam. This was a task requiring extreme care as traces of asbestos were found in the paintwork.

Below: June 2018 with the towers completely encased in a complex array of scaffolding.

Repainting the towers in June 2018. The discovery of asbestos in some of the earlier paint led to the bridge being encased in protective tents, the technicians inside working in reduced air pressure to ensure that no particles were allowed to escape into the environment.

engineering solutions before the residents of Rochefort or Echillais saw any evidence of activity on the site.

The residents were advised of the likely noise and disruption which the project would impose upon them, and in fact they endured it for longer than anticipated as severe gales and storms in the winter of 2017–18 caused considerable delays to the paint removal process.

Months of work preceded the removal of the 1933 beam – additional tensioned cables were fixed between the towers and down to the anchorages to maintain the integrity of the towers themselves as the loading on them was progressively removed. To facilitate that work, two temporary walkways were slung from the cables.

Both towers were encased in scaffolding, fitted with elevators and with temporary platforms constructed on top of them to aid the slinging of the additional support cables. Only after all those preparations were complete could the beam be cut into sections and, in late summer 2017, lowered on to barges waiting below.

At each stage of the removal, the tensions on the temporary support cables had to be adjusted to maintain structural integrity as the load they were carrying changed, and even in a light breeze, these cables hummed and pinged. All the pre-planning and computer modelling paid off handsomely, as the dismantling of the beam was completed without a hitch.

Once the asbestos in the paint had been successfully removed, some of the metalwork of the towers was found to be in very poor condition and had to be replaced.

A study of the several layers of paint revealed that, rather than the grey garb the bridge has worn for decades, it had originally been painted black, and in the cause of authenticity, the decision was made to restore the original colour – not seen since before the 1933 refurbishment. With modern paints applied, it should not require repainting for many years to come.

One of the pleasing facts about the reconstruction is that the project, being undertaken on behalf of the French Ministry of Culture and Communication and the local authority of the Rochefort-Océan region, was entrusted to Baudin–Chateauneuf, a company which has very strong associations with the bridge's original builders.

Left: The full length of the new 300 ton beam was temporarily bolted together on a closed section of road on the Echillais shore prior to being lifted into place, section by section, in autumn 2018. The movable blue structure was a temporary aid to the construction engineers and painters.

Below left and right: When the bridge was last restored it was felt necessary to infill the cable tunnels with concrete to support the additional weight of the plated beam. At the same time, additional cables were introduced, embedded in a 250 ton concrete addition to the rear of the cable anchorages. All that concrete has now been removed.

Set up in Chateauneuf-sur-Loire in 1919 and a direct successor to Arnodin's hugely productive enterprise in the town, the Baudin–Chateauneuf company was established by the entrepreneur Basile Baudin and the engineer Georges Camille Imbault, onetime assistant to Ferdinand Arnodin – for whom he supervised the building of the Newport transporter bridge and, later, was Chief Engineer to the Cleveland Bridge & Engineering Company who later designed the Middlesbrough transporter bridge (qv). Baudin–Chateauneuf still occupies part of Arnodin's original foundry in Chateauneuf-sur-Loire.

The *Architecte en Chef de Monument Historiques* was Philippe Villeneuve, and Lyon-based Artcad were the structural engineers for the project. Baudin–Chateauneuf and Artcad are now restoring an historic suspension bridge on the Loire.

Perhaps a reflection both of Arnodin's ingenious design for the Rochefort bridge, and the importance today's authorities place on health and safety, it will have taken nine months longer to restore the bridge than it took to build it in the first place – that was 27 months between 1898 and 1900.

But a great deal of that time has been taken in preparing the site, and undertaking the laborious task of undoing the ill-advised restorations of the past. Indeed, a substantial proportion of the €30M budget has gone on the temporary infrastructure necessary to facilitate that work.

The result will be a bridge looking almost exactly as it did when first opened. Up close there will be a number of slight modifications to rectify issues in Arnodin's original design – which will be largely invisible – designed to ensure the restored bridge has a long and sound future.

In the cause of authenticity, wherever possible the techniques used during the restoration mirror the methods of a century and more ago. Thus the sections of the new beam, like Arnodin's original, were hot-riveted together

Opposite above: Additional cables in place in May 2017 to support the towers before the removal of the 1933 beam.

Opposite middle left: The anchorages for the main cables at the top of the Rochefort tower, 66.25 metres above the river. The blue framework is a temporary platform to support cranes during the reconstruction.

Opposite middle right: The old saddles at the top of the towers, photographed in May 2017.

Opposite below: June 2018, the new saddles recently delivered to the site, ready for the installation of the new main deck beam.

The compression washers to relieve stresses will be invisible from the ground once the beam is in place.

Right: A view through the roof of the winch-house on the Rochefort shore, showing the cable reel. The reel is constantly open to the elements.

Below: From a specially-constructed viewing area near the site of the approach road to the former lift bridge in 2018, the massive scale of the new beam can be appreciated.

once they have been lifted into place. At the time the photographs on pages 00 and 00 were taken, those sections had been carefully aligned and temporarily bolted together on the ground.

The inherent flaw in Arnodin's original design – which caused the cracks to appear in the late 1920s – was that as the carriage crossed beneath the main beam, its weight, together with the increasing load of the vehicles being carried, imposed considerable stress on the fixed cross-braces, especially towards mid-channel, eventually resulting in fractures.

While that problem was effectively resolved by their replacement by the plated sides of the new beam in 1933, restoring the bridge today back to Arnodin's original configuration could have re-introduced it.

The neat solution to that challenge was to fit the braces with compression washers to relieve those stresses. The washers have an design life of at least 25 years before needing replacement.

The bridge's long-term future is now assured, and it will surely play a significant role in the forthcoming application, along with all the world's other surviving transporter bridges, to join Spain's Portugalete bridge on UNESCO's World Heritage Site lists.

Looking along the new beam – which replicates Arnodin's original design – as work progresses through 2019 on returning the bridge to service.

Like Arnodin's original design, the new beam is supported by a combination of vertical suspension cables and stay cables radiating from the tops of the towers.

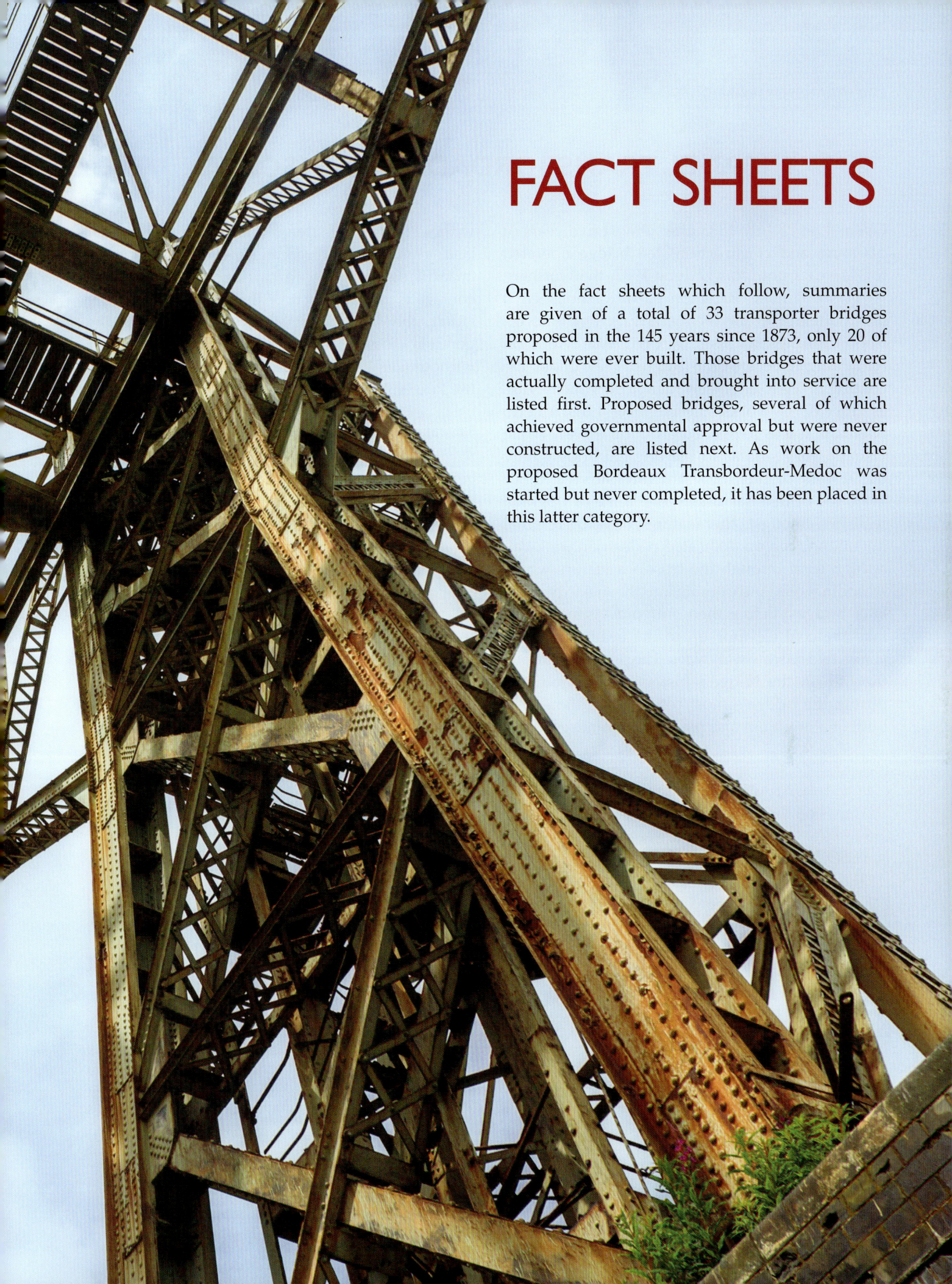

FACT SHEETS

On the fact sheets which follow, summaries are given of a total of 33 transporter bridges proposed in the 145 years since 1873, only 20 of which were ever built. Those bridges that were actually completed and brought into service are listed first. Proposed bridges, several of which achieved governmental approval but were never constructed, are listed next. As work on the proposed Bordeaux Transbordeur-Medoc was started but never completed, it has been placed in this latter category.

1893: EL TRANSBORDADOR DE VISCAYA

Opened on 28 July 1893 after almost five years of planning and construction, the bridge operated for 44 years until, on 17 June 1937, the Northern Army's Engineers Battalion partly demolished it in an attempt to hinder the advance of Franco's troops during the Spanish Civil War. Work on rebuilding it began on 5 August 1939 under the control of civil engineer Juan José Aracil with, as director of works, the engineer Luis Alberto Nieulant. It was re-opened on 19 June 1941. While visually very similar to the original bridge, only the towers of the 1893 structure survive. Since 1995, the bridge has undergone a major upgrade realising, amongst other features, Palacio's original intention to have a dedicated pedestrian walkway at high level through the truss – a feature which Arnodin would incorporate into the majority of his later bridges. Before then, intrepid visitors could take the coal elevator up to the beam and walk across two narrow walkways. The present gondola, the fifth in the bridge's history, was introduced during the bridge's most recent restoration. It is built of aluminium and plastic to minimise weight without sacrificing strength. Winch drive has now been replaced by direct electric motor drive on an articulated overhead carriage, ensuring good contact is retained across the entire span. Fully laden, the beam deviates around 25cm as the gondola crosses. As the only way across the river without a lengthy detour over a high-level bridge, the Puente Viscaya operates 24 hours a day.

Illustration: Looking north towards the bridge. Photographed by the author, May 2017.

LOCATION: Across the mouth of the Nervión River between Portugalete on the west shore and Las Arenas on the east, 16km north of Bilbao, Spain.
STATUS: Operational and in regular service.
ALSO KNOWN AS: Puente Bizkaia, Puente Colgante.
DESIGNED BY: Alberto Palacio.
ENGINEERED BY: Ferdinand Arnodin.
OPENED: 1893.
TYPE: Suspension bridge. The uncradled truss was originally supported by both vertical suspension cables and cable stays. Since rebuilding, supported by suspension cables only. Both splayed side anchorages and anchorages behind each tower.
DISTANCE BETWEEN TOWERS: 160 metres.
HEIGHT OF TOWERS: 61 metres.
CLEARANCE ABOVE HIGH WATER LEVEL: 45 metres.
MOTIVE POWER: Originally steam engines mounted at the Las Arenas end of the main beam, replaced in 1901 by steam-powered generator at lower level of Las Arenas tower, and electric winches. Since 1999 drive comes from 12 electric motors in the carriage bogies.
GONDOLA SIZE: 25 metres x 12 metres (current).
GONDOLA CAPACITY: The original gondola had a capacity of 150 passengers. The current gondola can carry 6 cars and 200 passengers. 40 tons total moving load.

1898: PONT TRANSBORDEUR BIZERTE/BREST

The world's second 'Pont Transbordeur' was erected across the mouth of the port of Bizerte or Viscaya in Tunisia, then controlled by the French. It was Arnodin's first 'solo' design and, with a span of 109 metres, was much narrower than the Viscaya Bridge which, in most respects, it resembled. As with all his later transporter bridges, the metalwork was prefabricated at his works at Chateauneuf-sur-Loire and then transported by train and ship to the site for re-assembly. Work on the bridge – a hybrid structure using both vertical suspension cables and cable stays took just under two years – started in 1897, was complete before the end of 1898. In the 1898 volume of *Memoires et compte-rendu des traveaux*, the journal of the Société des Ingenieurs Civil de France, Arnodin's gift of photographs of the bridge is noted. The bridge was dismantled after only nine years to make way for a widening of the harbour mouth to 200 metres – almost twice the bridge's span – to accommodate larger vessels. The component parts lay adjacent to the harbour for some months before being shipped to France and re-erected across the entrance to

the inner harbour at the 'Porte Militaire' in the Brittany port of Brest where it continued in use until damaged by Allied bombing in 1944. It was demolished in 1947. In 2014, a suggestion was made to build a new transporter bridge across the mouth of the port, primarily as a tourist attraction, its suggested design being a modern version of Arnodin's cantilevered suspension design at Nantes and Marseille.

Illustration: The rebuilt bridge at Brest, from a postcard published around 1903.

LOCATION: Bizerte, Tunisia, 1898–1909; dismantled and rebuilt at Brest, Brittany, France, 1909.
STATUS: Damaged 1944, demolished 1947.
ALSO KNOWN AS:
DESIGNED BY: Ferdinand Arnodin.
ENGINEERED BY: Ferdinand Arnodin.
OPENED: 1898.
TYPE: The uncradled truss was supported by a hybrid of vertical suspension cables and cable stays in the same manner as the original Viscaya Bridge.
DISTANCE BETWEEN TOWERS: 109 metres.
HEIGHT OF TOWERS: 58 metres.
CLEARANCE ABOVE HIGH WATER LEVEL: 45 metres.
MOTIVE POWER: Originally driven by steam winches, later converted to electricity.
GONDOLA SIZE: 12 metres x 8 metres.
GONDOLA CAPACITY: Reported (at Brest) as 100–200 passengers and one vehicle, estimated 40 tons total moving load. However, the engineer at Viscaya had reported in 1906 that 'according to carefully prepared statistics, it averages 272 journeys per day, carrying 4,220 passengers, 1,500 animals and 503 vehicles'. Those two sets of figures cannot be reconciled.

1899: PONT TRANSBORDEUR DE ROUEN

The third bridge to be completed by Frederick Arnodin, the transbordeur at Rouen followed the design used for the Viscaya bridge, albeit with slightly taller pylons and greater clearance above river level. It was also faster and could offer a crossing time of just 45 seconds, and its daily capacity was said to be 240 crossings, carrying 300 vehicles and up to 10,000 passengers. For passengers, there was a choice of first or second class travel – second class being open to the elements – and the bridge could carry anything from a single horse with a two-wheeled rig up to large and heavy four-horse vehicles. Construction started in 1895 and the bridge opened for traffic early in 1899.

The transporter bridge, together with the Pont Boieldieu and the Viaduc d'Eauplet were all destroyed by the French in June 1940 in an attempt to slow down the invading German troops. It was never rebuilt.

Illustration: Postcard published by Léon & Lévy around 1903.

35 ROUEN. — Le Pont Transbordeur, Hauteur des Pylones, 60 mètres. – LL.

LOCATION: Across the River Seine in Rouen, Normandy, France.
STATUS: Destroyed, 1940.
ALSO KNOWN AS:
DESIGNED BY: Ferdinand Arnodin.
ENGINEERED BY: Ferdinand Arnodin.
OPENED: 1899.
TYPE: The uncradled stiffening beam was supported by a hybrid of suspension cables and cable stays in the same manner as the original Viscaya Bridge.
DISTANCE BETWEEN TOWERS: 143 metres.
HEIGHT OF TOWERS: 67 metres.
CLEARANCE ABOVE HIGH WATER LEVEL: 51 metres
MOTIVE POWER: Originally driven by steam winches, later changed to electricity.
GONDOLA SIZE: 13 metres x 10 metres.
GONDOLA CAPACITY: 200 passengers and 6 vehicles. 45 tons total moving load.

1900: PONT TRANSBORDEUR DE ROCHEFORT

Work on the transporter bridge across the River Charente at Rochefort-sur-Mer – Arnodin's first design to use an overshooting beam – began in spring 1898 and was completed by July 1900. The gondola is suspended from a 24-wheeled travelling frame. Control of the gondola was initially in the hands of the motor-man in the winch-house sited behind the Rochefort tower. Initially the crossing time was 75 seconds, but in the years before the current closure, it was usually around four minutes. After just 32 years of operation, cracks were found in the side trusses of the stiffening beam, requiring the bridge to be closed for more than a year and the beam reinforced. As a result of that work, the carrying capacity was increased to 16 tons. Further restoration work between 1980 and 1994 ensured the bridge's continued use as a tourist attraction but, in November 2015, the start of a new three-year €30M restoration programme was announced to replace much of the 1933 metalwork and restore the bridge to its 1900 appearance. Work on that started early in 2016. The bridge is out of service during the rebuild, and reopening is scheduled for late 2020. The highly informative 'Maison de Transbordeur' visitor centre has remained open during the restoration, telling the story of the bridge and including a video 'virtual crossing' of the river. While the bridge was closed, a small ferry plied between the two shores 200 metres downstream from the bridge.

Illustration: Photographed by the author during the early stages of restoration, May 2017.

LOCATION: Rochefort-sur-Mer, Charente-Maritime, France.

STATUS: Currently undergoing major restoration, re-opening late 2020. Visitor Centre still open.

ALSO KNOWN AS: Le Pont Transbordeur Rochefort–Martrou, Le Pont de Martrou, Le Pont Transbordeur Rochefort–Echillais.

DESIGNED BY: Ferdinand Arnodin.

ENGINEERED BY: Ferdinand Arnodin.

OPENED: 1900.

TYPE: The uncradled overshooting beam was originally supported by a combination of vertical suspension cables and cable stays. From 1933 until 2017 it just had vertical suspension cables. A new beam was re-installed in 2019, and the cable stays re-instated.

DISTANCE BETWEEN TOWERS: 140 metres.

HEIGHT OF TOWERS: 66.25 metres.

CLEARANCE ABOVE HIGH WATER LEVEL: 47 metres.

MOTIVE POWER: Originally driven by steam-powered winches on the Rochefort shore, converted to electricity in 1927.

GONDOLA SIZE: 14.6 metres x 11.6 metres.

GONDOLA CAPACITY: Originally 200 passengers or a combination of passengers and vehicles to a maximum capacity of 14 tons. Capacity raised to 16 tons or 12 motor cars in 1934. Since 1994, capacity limited to 100 passengers & cyclists.

1903: PONT TRANSBORDEUR DE NANTES

The first of two cantilevered cable-stayed bridges designed by Arnodin, construction of the Pont Transbordeur at Nantes caused some considerable interest and its progress was photographed by leading postcard publishers Léon & Lévy. Not only was it the first time Arnodin had employed such a design, it was also the first time he had used direct drive from DC motors in the overhead travelling frame. The project was first discussed not long after the opening of the Viscaya bridge and was eventually given approval in 1898. Construction took less than two years, beginning early in 1902 and opening in November 1903. Arnodin's own works at Chateauneuf-sur-Loire manufactured much of the steelwork. The 46-ton central section of the cross-beam was raised in to position on 3 August 1903, with the completed bridge being load-tested just a few weeks later. During that load test, the gondola successfully carried 85 tons of stone across the river, almost 50 per cent more than its design capacity. In passenger-carrying service, the total passage time was one minute.

It was one of only two transporters in France to survive the Second World War, but the traffic using the bridge reduced considerably and despite protests the service was withdrawn in January 1955 and the bridge dismantled in 1958. Only the stone piers for the bridge survive, built in to the quayside.

'Projet Jules Verne' is a current proposal for the building of a modern transporter bridge as a tourist attraction across approximately the same stretch of the river.

Illustration: Postcard published by Léon & Lévy around 1904.

165 NANTES. — Le Pont Transbordeur. — LL.

LOCATION: Across the Madelaine branch of the River Loire in Nantes, Brittany, France.
STATUS: Dismantled 1958.
ALSO KNOWN AS:
DESIGNED BY: Ferdinand Arnodin.
ENGINEERED BY: Ferdinand Arnodin.
OPENED: 1903.
TYPE: Uncradled cable-stayed suspension bridge with overshooting beam and cantilevered central section, stabilised with vertical anchorages beneath each end of the beam.
DISTANCE BETWEEN TOWERS: 141 metres.
HEIGHT OF TOWERS: 75 metres.
CLEARANCE ABOVE HIGH WATER LEVEL: 50 metres.
MOTIVE POWER: DC Electric motors on the overhead travelling frame.
GONDOLA SIZE: 10 metres x 12 metres.
GONDOLA CAPACITY: Exact load not specified, but a maximum carrying capacity of 60 tons when in service has been reported.

1905: DULUTH AERIAL FERRY BRIDGE

The Duluth bridge was America's only transporter, and a short-lived one at that. Opened in 1905, it was modified to operate as a vertical lift bridge in 1929, using the original towers extended upwards. Thomas McGilvray got his inspiration from Rouen's 'transbordeur' which he had visited just after it was completed and, despite looking very different to Arnodin's 1899 bridge, the Duluth bridge's drive mechanism was remarkably similar, but with the cable winches built into the sides of the gondola rather than on top as at Rouen. In the original design, there were intended to be overhead power lines on the gondola to enable electric streetcars to run on and off under their own traction, but this innovation never seems to have been incorporated into the final bridge. The crossing time was around one minute. McGilvray's original concept was engineered by eminent structural engineer Claude Allen Porter Turner of Lincoln, Rhode Island. When the decision was made to convert it to a vertical lift bridge, Porter Turner submitted his own engineering solution, but his design was rejected and the work was carried out by Harrington, Howard & Ash of Kansas City. The lift bridge still operates today and is one of the town's defining structures.

Illustration: Postcard published by the Kropp Company of Milwaukee, Wisconsin, at the time of the bridge's opening in 1905.

LOCATION: Across the Duluth Ship Canal at the mouth of Duluth Harbour, Minnesota, USA.
STATUS: Converted to a lift bridge in 1929.
ALSO KNOWN AS:
DESIGNED BY: Thomas McGilvray.
ENGINEERED BY: Claude Allen Porter Turner.
OPENED: 1905.
TYPE: Rigid self-supporting lattice girder.
DISTANCE BETWEEN TOWERS: 121 metres.
HEIGHT OF TOWERS: 50 metres.
CLEARANCE ABOVE HIGH WATER LEVEL: 41 metres.
MOTIVE POWER: Electric winch drums in the sides of the gondola.
GONDOLA SIZE: 16 metres x 10 metres.
GONDOLA CAPACITY: Original design envisaged 350 people plus wagons, electric tramcars and cars, 54 tons total load.

1905: PONT TRANSBORDEUR DE MARSEILLE

The 'Pont a Transbordeur' across the mouth of the harbour between Fort Saint John and Fort Saint Nicolas in Marseille – the fifth of Arnodin's French transporter bridges – was constructed in 19 months and opened on 15 December 1905. Arnodin had negotiated a 99-year concession to build and operate the bridge. It was the second and larger of his two cantilevered bridges, requiring a slightly wider span than the Nantes transporter, its towers being proportionately taller as a result. The gondola, however, was very similar in design and size to that used at Nantes. As with the Nantes bridge, he employed a direct DC electric drive system on the bogies of the travelling frame. At the opposite ends of the beam, the bridge had its own 'Buffet-Restaurants' and souvenir shops where tourists could buy postcards and other mementos of their experience after 'walking the beam'. The bridge might easily have been demolished when just a decade old, as its steel – almost 1,200 tons – was about to be requisitioned for war use when the First World War came to an end. It did not fare so well in the Second World War, and on 22 August 1944 the Germans decided to blow it up to block the port entrance, but only the north tower fell. The rest of the bridge was brought down by explosives on 1 September 1945. Since 2014, interest has been growing in a plan to build a new transporter, primarily as a tourist attraction, crossing le Vieux Port at approximately the same point. The design of the proposed new bridge – by architect Paul Poirier – would use a full-length stiffening beam, without cantilevering.

Illustration: From a postcard c.1920.

LOCATION: Across the harbour mouth, Marseille, France.
STATUS: Destroyed, 1944–45.
ALSO KNOWN AS:
DESIGNED BY: Ferdinand Arnodin.
ENGINEERED BY: Ferdinand Arnodin.
OPENED: 1905.
TYPE: Uncradled cable-stayed suspension bridge with overshooting beam and cantilevered central section, stabilised with vertical anchorages beneath each end of the beam.
DISTANCE BETWEEN TOWERS: 165 metres.
HEIGHT OF TOWERS: 84 metres.
CLEARANCE ABOVE HIGH WATER LEVEL: 50 metres.
MOTIVE POWER: Electric winches in winch-house at one end of stiffening beam.
GONDOLA SIZE: 10 metres x 12 metres.
GONDOLA CAPACITY: About 100 passengers and four vehicles, or 200 passengers and one vehicle; a maximum moving load of 60 tons has been suggested.

1905: WIDNES & RUNCORN TRANSPORTER BRIDGE

The Widnes & Runcorn Bridge Company, established in 1899, was granted parliamentary approval for the construction of the bridge in 1901, and work started before the end of that year. The bridge was opened in May 1905 by Sir John Brunner, deputising for King Edward VII who was indisposed at the time. It cost £130,000 to build. It is estimated that the gondola, in its heyday, could be required to make as many as 160 crossings per day in order to meet the growing demand from vehicle traffic – way beyond what it was ever designed to do. The bridge was revolutionary in its day, generating its own electricity to drive the Mather & Platt DC motors that were built in to the overhead carriage bogies. This proved a source of operational problems as the stiffening girder flexed under high load giving the overhead carriage intermittent contact with the live rail. The direct drive system was replaced by cable-drive in 1913 with the erection of a winch-house on the Widnes side. As traffic volume increased the transporter's replacement by a fixed bridge was inevitable, and work started on the present-day road bridge in 1956. It eventually opened in 1961, after which the transporter was immediately taken out of service.

Over the following two years, the bridge was demolished, but the approach roads to it survive, as does one of the 1905 buildings on the Widnes shore, now Grade II* listed, and the bridge company's offices a few yards further up Mersey Road, Widnes.

Illustration: Benbow series tinted postcard of the completed bridge published at the time of its opening in 1905.

LOCATION: Spanning the River Mersey and Manchester Ship Canal, between Runcorn, Cheshire, and Widnes, Lancashire, UK.

STATUS: Dismantled 1961–62.

ALSO KNOWN AS: Runcorn Transporter Bridge, Widnes Transporter Bridge, Runcorn and Widnes Transporter Bridge.

DESIGNED BY: John J. Webster and John T. Wood.

ENGINEERED BY: Arrol Brothers Bridge & Roof Company, Germiston Works, Glasgow.

OPENED: 1905.

TYPE: Suspension bridge with cradled overshooting beam. Towers stabilised by anchorages behind each tower.

DISTANCE BETWEEN TOWERS: 304 metres.

HEIGHT OF TOWERS: 58 metres.

CLEARANCE ABOVE HIGH WATER LEVEL: 25 metres.

MOTIVE POWER: Initially, electricity generated by two 75hp Crossley gas engines supplying power to DC electric motors in the overhead carriage bogies. Replaced by cable drive in 1913.

GONDOLA SIZE: 16.8 metres x 7.3 metres.

GONDOLA CAPACITY: As built, 300 passengers and 'four two-horse loaded lurries'.

1906: NEWPORT TRANSPORTER BRIDGE

The only bridge to be built outside France to a design by Frederick Arnodin, the Newport bridge is now the longest transporter bridge still standing anywhere in the world – while the main stiffening beam is not as long as the 1911 cantilevered Tees bridge, the span is 17 metres greater.

Borough Engineer Robert Haynes had been impressed by the Rouen 'transbordeur' during a visit to the French city in 1899, and as a result Newport City Council directly engaged Ferdinand Arnodin to build the bridge over the Usk.

An enabling bill was laid before Parliament in 1900, work started on site in 1902 and the bridge was opened by the 1st Viscount Tredegar in September 1906. The total cost was £98,000.

The project, reportedly, endured some challenging times. With the design work being carried out at Arnodin's base in Chateauneuf-sur-Loire, and manufacture and erection taking place in Newport, there were apparently instances of confusion over the interpretation of metric specifications in a country where everything was in imperial sizes.

The bridge operated for almost 80 years before being shut down for safety reasons in 1985. It re-opened a decade later after a major restoration, but just 13 years after that further major refurbishment was necessary and it was closed again from 2008 until 2010. Efforts are afoot to get it listed on the UNESCO World Heritage list, along with the other surviving transporter bridges.

Illustration: Photographed by the author, 2016.

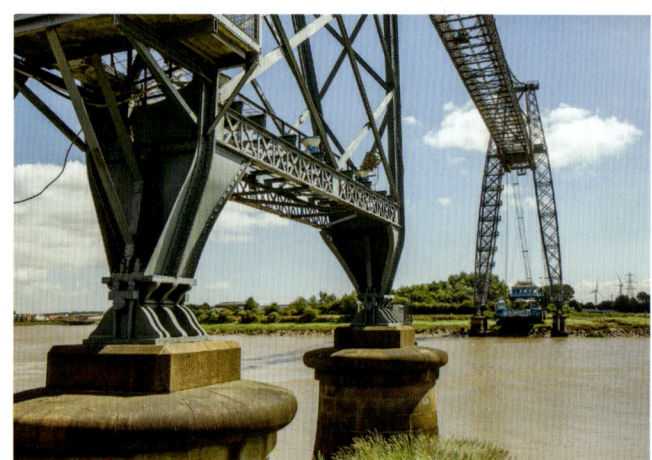

LOCATION: The River Usk, Newport, South Wales, UK.

STATUS: Seasonal operation only, currently runs Wednesday to Sunday, May to October.

ALSO KNOWN AS:

DESIGNED BY: Ferdinand Arnodin.

ENGINEERED BY: John P. Thorne for Alfred Thorne & Sons; Georges Camille Imbault for the Cleveland Bridge & Engineering Company; and Robert Henry Haynes, Newport Borough Engineer.

OPENED: 1906.

TYPE: The uncradled main overshooting beam is supported by a hybrid of suspension cables and cable stays in the same manner as the Rochefort–Martrou Bridge.

DISTANCE BETWEEN TOWERS: 197 metres.

HEIGHT OF TOWERS: 74 metres.

CLEARANCE ABOVE HIGH WATER LEVEL: 54 metres.

MOTIVE POWER: Two 35hp (25kw) electric winches in winch-house above and behind the eastern landing stage of the bridge.

GONDOLA SIZE: 12 metres x 9 metres.

GONDOLA CAPACITY: 100 passengers and 3 vehicles. 45 tons total moving load.

1905: WARRINGTON CROSFIELD TRANSPORTER NO.1

The first Warrington Transporter Bridge was brought into service in 1908, the first of only two ever built in Britain to which the public had no access. The second one was built just a few hundred yards away less than a decade later. Crosfield No.1 had a relatively low carrying capacity at just 2.5 tons which, at the design stage, had been considered more than adequate. As a result, the towers and stiffening cross-beam could be of relatively lightweight construction. One tower was built on solid foundations at factory yard level, while the other was sited on the roof of one of the company's warehouses. Stability was achieved by the use of guy ropes/cables attached to the pylon wings at cross-beam level and splayed out behind each tower. The original concept was by Crosfield's resident engineer James Newhall and construction was entrusted to Thomas Piggott & Company's Atlas Works in Spring Hill, Birmingham, who were already established builders of steel lattice bridges. Crosfield's bridge was one of only three ever built to use DC motors directly driving the bogies of the overhead carriage, a decision made after inspecting the Widnes–Runcorn bridge which had been completed in 1905. By 1913, that system had been abandoned on the Widnes and Runcorn bridge – being replaced by cable and winch drive – leaving Crosfield's bridge the only one in Britain using this drive method. The transporter bridge continued to give service long after the second and much stronger bridge was commissioned in 1916 and it was demolished in the 1960s. Just when it had been withdrawn from service is unknown.

Illustration: Published in *The Engineer* 3 April 1908.

LOCATION: Bank Quay, Warrington, Cheshire, UK.
STATUS: Demolished in the 1960s.
ALSO KNOWN AS: Crosfield Transporter Bridge No.1.
DESIGNED BY: Thomas Piggott & Company of Birmingham.
ENGINEERED BY: James Newall, and Thomas Piggott & Co.
OPENED: 1908.
TYPE: The uncradled twin-girder beam was supported by two cables suspended between single towers, one bedded into the ground, the other seated on the top of one of the factory buildings. Ground-anchored lateral support cables radiated from the tower 'wings'.
DISTANCE BETWEEN TOWERS: 78 metres.
HEIGHT OF TOWERS: 60 metres (estimate).
CLEARANCE ABOVE HIGH WATER LEVEL: 50 metres.
MOTIVE POWER: DC Electric motors directly driving the overhead carriage bogies.
GONDOLA SIZE: Unknown.
GONDOLA CAPACITY: 2.5 tons load capacity.

1909: SCHWEBEFAHRE OSTEN–HEMMOOR

Standing in the Cuxhaven District of Lower Saxony, the Osten–Hemmoor Bridge, the first of the three transporter bridges to be built in Germany, was an early example of a public/private funding partnership. The local community, a leading bank, and Claus Drewes the former ferry operator were all founder investors in the project. The bridge's lightweight design, with a rigid suspension frame rather than cables, had no precedent, although it exhibits several design similarities to ideas developed at Duluth, Minnesota and, in using direct electric motor drive to its four-wheeled overhead bogie, it operated on a simplified version of the drive system installed on the Widnes–Runcorn Bridge over the Mersey four years earlier. The steelwork for the bridge was manufactured by Fundamente der Maschinenfabrik Augsburg-Nürnberg, one of the component companies of the MAN Group, and the electrical systems were designed and installed by Allgemeine Elektrizitäts-Gesellschaft (AEG) of Berlin. The asymmetric design of the bridge's superstructure was dictated by the nature of the landscape and the necessity for sound foundations. The total construction cost was around 280,000 Deutschmarks – around £28,000 at the time – a huge cost for a small community of 600 inhabitants to bear. It was, however, 40 per cent cheaper than building a swing bridge. The official opening ceremony took place on 1 October 1909.

Illustration: Aerial photograph © Petra Klawikowski, reproduced under Creative Commons license.

LOCATION: Bridging the canalised River Oste near Cuxhaven, Lower Saxony, Germany.
STATUS: In regular use until 1974, it is now operated seasonally as a visitor attraction.
ALSO KNOWN AS: Osten Hover Ferry.
DESIGNED BY: Freiherr von Schröder.
ENGINEERED BY: Max Pinette.
OPENED: 1909.
TYPE: Lightweight truss.
DISTANCE BETWEEN TOWERS: 79 metres.
HEIGHT OF TOWERS: 38 metres.
CLEARANCE ABOVE HIGH WATER LEVEL: 21 metres.
MOTIVE POWER: Direct drive electric motors on the overhead four-wheeled bogie. Power supply changed from DC to three-phase in the 1920s.
GONDOLA SIZE: 16 metres × 4.30 metres
GONDOLA CAPACITY: 25 passengers and 2 vehicles, or 6 cars, or a full-size coach. Maximum gondola payload, 18 tons.

1910: SCHWEBEFAHRE KEIL

The second of Germany's three transporter bridges spanned the entrance to the naval harbour at Kiel and was built by 'Gutehoffnungshütte, Aktienverein für Bergbau und Hüttenbetrieb' – a mining and industrial conglomerate based in Oberhausen and usually referred to as GHH – to a design by Georg Franzius, Director of Kiel's Imperial German shipyard. It would be Franzius's last major project in an eminent engineering career. The earliest proposal to build a transporter bridge across the mouth of the dockyard dates from 1902, but the project was not approved until 1908, construction of the bridge being complete by late 1909 or early 1910. The design of the bridge has many similarities to Palacio and Arnodin's 1893 Viscaya bridge, except that the stiffening beam was entirely supported by vertical suspension cables. The gondola was fitted with railway track and could accommodate two goods wagons. As the coal-yard was on one side of the docks, it can be presumed that much of the traffic using the bridge was concerned with the distribution of coal around the naval dockyard. The bridge had a surprisingly short working life, being dismantled in 1923 to facilitate the widening of the entrance to the dockyard and harbour. Despite being the subject of many contemporary postcards, its history remains relatively obscure.

Illustration: From a postcard published by, amongst others, Verlag Hermann Edlefsen around 1911 and postmarked 1913. At least six different publishers produced this same card, in both sepia and tinted versions.

LOCATION: Naval Harbour Kiel, Schleswig-Holstein, Germany.
STATUS: Demolished 1923 to make way for harbour improvements.
ALSO KNOWN AS:
DESIGNED BY: Georg Ludwig Franzius, Kiel's Marine Port Engineer.
ENGINEERED BY: Georg Franzius and Gutehoffnungshütte, Aktienverein für Bergbau und Hüttenbetrieb.
OPENED: 1910.
TYPE: Suspension bridge with uncradled truss. Bridge anchored by cables behind each tower.
DISTANCE BETWEEN TOWERS: 128 metres.
HEIGHT OF TOWERS: 70 metres.
CLEARANCE ABOVE HIGH WATER LEVEL: 49 metres.
MOTIVE POWER: Winches driven by a steam engine located in winch house behind one of the towers.
GONDOLA SIZE: Unknown.
GONDOLA CAPACITY: 2 railway goods wagons and/or an unspecified number of passengers.

1911: TEES TRANSPORTER BRIDGE (3)

It took four separate proposals and 38 years before a transporter bridge finally opened over the Tees between Middlesbrough and Port Clarence. The bridge which was finally opened bore more than a passing resemblance to Charles Smith's 1873 design. A proposal by Ferdinand Arnodin – believed to have been for a more conventional cable-slung transporter – was rejected, although Arnodin remained a consultant on the project. Considerable opposition to the bridge by some local councillors was eventually overcome, delaying the project by several years. Designed by Georges Imbault – Arnodin's former assistant – Chief Engineer for the Cleveland Bridge & Engineering Company, but built by Sir William Arrol & Company of Dalmarnock, Glasgow, the bridge is now *the* iconic symbol of Teesside. The steelwork was prefabricated in Glasgow and transported to the site, enabling rapid assembly once the foundation piers were in place. The project time was 27 months, but experienced significant cost overruns, leading to suggestions that Arrols had deliberately under-estimated the costings in their successful tender.

Others have suggested that had the real final cost been known in advance, the entire project might never have got council support. However, 107 years after completion, the bridge is still working, although now only relatively lightly used since a high level bridge eased traffic loads.

Photograph by the author, 2017.

LOCATION: Across the River Tees between Middlesbrough and Port Clarence, Teesside, UK.
STATUS: Still in use six days a week. Closed Sundays.
ALSO KNOWN AS: Middlesbrough Transporter Bridge, the 'Tranny'.
DESIGNED BY: Georges Camille Imbault and Alfred J. Collin for the Cleveland Bridge & Engineering Company.
ENGINEERED BY: Sir William Arrol & Company, Chief Engineer Robert Anderson.
OPENED: 1911.
TYPE: Cantilevered bridge with vertical anchorages.
DISTANCE BETWEEN TOWERS: 180 metres.
HEIGHT OF TOWERS: 68 metres.
CLEARANCE ABOVE HIGH WATER LEVEL: 48.7 metres.
MOTIVE POWER: Originally Westinghouse 500 volt DC motors powering a winch through a geared cog drive. Since 2000, three-phase motors from Enigma SX power the same winch.
GONDOLA SIZE: Original design size 11.9 metres x 13.4 metres. Current gondola 10.7 metres x 14.6 metres.
GONDOLA CAPACITY: Original gondola: 600 passengers; current gondola, a combination of passengers and up to 9 vehicles.

1913: SCHWEBEFAHRE RENDSBURGER

The 'Rendsburger Schwebefähre' is a unique structure – the only dual-function rail/transporter bridge ever built. The bridge – designed by the engineer Friedrich Voss, who had been appointed as head of the newly created bridge construction office in the German Imperial Office of Canals in 1908 – replaced an earlier railway swing bridge which was causing considerable interruption of canal traffic. The addition of a transporter system slung below the railway bridge provided easy road access between Osterrönfeld and Rendsburg. The bridge's rigid construction – necessary to carry the weight of trains – meant direct electric drive on to the transporter carriage's bogies was not beset by contact problems. Unlike many of the world's other transporters, where the gondola crosses just above the water level, at Rendsburg the clearance is eight metres. However, the gondola has proved to be somewhat accident-prone in recent years. It broke loose in a storm in January 1993 and was blown part way across the canal, colliding with a passing ship.

The transporter was later closed after the 1,598grt cargo ship *Evert Prahm* was in collision with the gondola, causing severe damage on the night of 8 January 2016, and there are currently no plans to reinstate it. Despite the established rule that the gondola should be held at one side of its traverse if there is an approaching ship within 800 metres – the crossing takes about 90 seconds – the collision took place in mid-canal, causing the gondola to be swept violently to one side and left suspended and stranded with passing ships having to navigate around it. Two people were slightly injured.

Illustration: Used under licence from Pixabay.

LOCATION: Across the Kiel Canal, Schleswig-Holstein, Germany.
STATUS: Normally operational, but out of service for repairs at the time of writing.
ALSO KNOWN AS:
DESIGNED BY: Friedrich Voss, Head of Bridge Department, German Imperial Canal Authority.
ENGINEERED BY: Friedrich Voss.
OPENED: 1913.
TYPE: Cantilevered lattice truss viaduct with transporter below railway track.
DISTANCE BETWEEN TOWERS: 140 metres.
HEIGHT OF TOWERS: 50 metres.
CLEARANCE ABOVE HIGH WATER LEVEL: 42 metres.
MOTIVE POWER: Direct electric drive to carriage bogies.
GONDOLA SIZE: 14 metres x 6 metres.
GONDOLA CAPACITY: 60 passengers and up to 6 vehicles.

1914: BUENOS AIRES PUENTE SÁENZ PENÃ

The first of the three transporter bridges built in Buenos Aires, the Puente Transbordador Sáenz Penã seems to have been the least photographed, and is certainly the one about which least is known. Contrary to what is stated in the UNESCO citation for the Viscaya Bridge in Spain, the three Buenos Aires bridges were not all built to the same design. Two of them, the Transbordador Sáenz Penã and the Transbordador Justo José de Urquiza, appear to have been almost identical in both size and design – simple, small, self-supporting and untethered structures. The bridges were manufactured by a French/German engineering consortium led by Pasquet et Cie, and shipped to Argentina for assembly on site. The gondola, a rigid steel structure with the driving cab on the roof, was suspended from the travelling frame – which ran on four pairs of bogies – by four rigid metal poles, stabilised by eight cross-wires to minimise lateral deflection. Some confusion exists over the date of this bridge's inauguration, with some sources saying it was opened in March or April 1913, others giving the following year. 1914 seems the more reliable date. The Puente Transbordador Sáenz Penã linked Garibaldi Street in Buenos Aires with what is now the Parque Logístico Sur in Avellaneda. It operated for 52 years before being closed and immediately demolished in 1965.

Illustration: From a 1930s postcard.

Buenos Aires — Vista panorámica tomada desde el puente
"Presidente Luis Saenz Peña" - Parte Sud

LOCATION: Across the Riachuelo-Matanza river, Buenos Aires, Argentina.
STATUS: Dismantled, 1965.
ALSO KNOWN AS: Puente Doctor Luis Saenz Peña.
DESIGNED BY: Unknown.
ENGINEERED BY: Pasquet et Cie.
OPENED: 1913 or 1914.
TYPE: Rigid self-supporting lattice-truss bridge.
DISTANCE BETWEEN TOWERS: Estimated at 60 metres.
HEIGHT OF TOWERS: Estimated at 46–48 metres.
CLEARANCE ABOVE HIGH WATER LEVEL: Estimated at 42–46 metres.
MOTIVE POWER: Electric winch drive, with the winch house sited on top of one of the towers.
GONDOLA SIZE: Estimated at 12 metres x 9 metres.
GONDOLA CAPACITY: Unspecified. According to some (unconfirmed) sources, the gondola may have been able to carry a tramcar.

1914: BUENOS AIRES PUENTE N. AVELLANEDA

The Puente Transbordador Nicolás Avellaneda was designed and built for the Ferrocaril del Sud – the Buenos Aires Great Southern Railway – one of four British-owned broad-gauge railways operating in Argentina. Contrary to a number of published descriptions, and although British-owned, the BAGSR had nothing to do with the Southern Railway in Britain – that company was not established until UK railway grouping in 1923. Approval for the bridge was granted on 25 September 1908 and construction took six years. According to plaques still fixed to the bridge, the metalwork was manufactured in the UK at The Earl of Dudley's Round Oak Steelworks in Brierley, West Midlands, (then in Staffordshire) and shipped to Argentina in sections for re-assembly. It crosses the Riachuelo-Matanza river linking the La Boca area with the port city of Avellaneda. Originally scheduled to open on 8 March 1914, that was postponed due to problems with the winch mechanisms and it finally opened on 30 May 1914, having cost a total of £100,000. Reliability issues beset the bridge in the 1930s prompting its replacement with a lift bridge, which opened in 1940. The bridge re-opened in 2018 as a tourist attraction after a major refurbishment, and with a new winch-house and new winding machinery installed.

Illustration: The bridge in 2017 with restoration nearing completion. Photograph courtesy of Dr Ron Callender.

LOCATION: Across the Riachuelo-Matanza river, Buenos Aires, Argentina.
STATUS: Unused since the 1960s, the bridge is now restored and returned to service as a tourist attraction.
ALSO KNOWN AS: Puente Transbordador de La Boca, Antiguo Puente Nicolás Avellaneda (after its original name was transferred to the 1940 road bridge), Transbordador del Riachuelo, Puente Colgante del Riachuelo.
DESIGNED BY: Santiago Podesta.
ENGINEERED BY: Steelwork by Earl of Dudley Steelworks, Brierley; on-site assembly by Construcciones Portuarias Buenos Aires.
OPENED: 1914.
TYPE: Self-supporting lattice truss bridge with rigid gondola suspension.
DISTANCE BETWEEN TOWERS: 77 metres.
HEIGHT OF TOWERS: 43 metres.
CLEARANCE ABOVE HIGH WATER LEVEL: 43 metres.
MOTIVE POWER: Electric winches.
GONDOLA SIZE: 12 metres x 8 metres.
GONDOLA CAPACITY: Unknown. Some sources claim it was fitted with tramlines, allowing the carriage of tramcars as well as road vehicles and pedestrians, but surviving contemporary photographs show no evidence of rails on the gondola deck. It was, however, clearly capable of carrying a significant weight.

1915: BUENOS AIRES PUENTE JOSE DE URQUIZA

Like the Puente Transbordador Sáenz Penã, which opened a year earlier, the bridge is believed to have been manufactured by a French/German engineering consortium led by Pasquet et Cie, and shipped to Argentina for assembly on site. The third and last of Buenos Aires' transporter bridges, it opened to traffic in March 1915, linking Deán Funes in Avellaneda with Avenida Regimento de Patricios in Barracas. Despite the simple construction of the gondola, both this bridge – and its 'twin' the Puente Transbordador Sáenz Penã – would have been able to transport a relatively heavy load. The gondola, a rigid steel structure with the driving cab on the roof, was suspended from the travelling frame – with ran on four pairs of bogies – by four suspended rigid metal poles, stabilised by eight cross-wires to minimise lateral deflection.

Illustration: From a photograph c.1915.

LOCATION: Across the Riachuelo-Matanza river, Buenos Aires, Argentina.
STATUS: Dismantled 1968.
ALSO KNOWN AS: Puente Transbordador Capitán General Justo José de Urquiza, Puente Transbordador Patricios.
DESIGNED BY: Unknown.
ENGINEERED BY: Pasquet et Cie.
OPENED: 1915.
TYPE: Rigid self-supporting lattice-truss bridge.
DISTANCE BETWEEN TOWERS: Estimated at 60 metres.
HEIGHT OF TOWERS: Estimated at 46–48 metres.
CLEARANCE ABOVE HIGH WATER LEVEL: Estimated at 42–44 metres.
MOTIVE POWER: Electric winch drive, with the winch house sited on top of one of the towers.
GONDOLA SIZE: Estimated at 12 metres x 9 metres.
GONDOLA CAPACITY: Unspecified. According to some (unconfirmed) sources, the gondola may have had been able to carry a tramcar.

1915: RIO DE JANIERO PONTE ALEXANDRINO

The Ponte Alexandrino, the only transporter to be erected in Brazil, was built to link the Central Naval Hospital and Arsenal on the Ilha das Cobras – Snake Island – with Rio de Janiero on the mainland. It was commissioned by Admiral Alexandrino Faria de Alencar, then Minister in charge of the Brazilian Navy, and was built by a German engineering consortium. It had several structural similarities to the much larger Widnes–Runcorn Bridge which had been completed a decade earlier, but used a rigid metal frame from which to suspend the gondola car, similar to those on the bridges at Duluth, Buenos Aires and Osten–Hemmoor. As originally built, and as at Widnes, drive was from DC motors on the overhead travelling frame, the electricity for which was generated by steam engines driving a dynamo in a powerhouse on the mainland side. At the time of its construction, the waterway between the island and the mainland was still considered to be an important shipping lane, ruling out the construction of a conventional bridge. And yet, just twenty years after its completion, the bridge was surplus to requirements and was dismantled in 1935. By that time, developments in shipping, and extensive redevelopment of the naval base and arsenal meant that the high clearance afforded by the transporter bridge was no longer necessary. A new low-level bridge had been built alongside it by 1930.

Illustration: From a postcard published c.1916.

RIO DE JANEIRO – PONTE ALEXANDRINO, DO ARSENAL Á ILHA DAS COBRAS

LOCATION: Linking Rio de Janeiro with Ilha das Cobras, Rio de Janeiro, Brazil.
STATUS: Dismantled, 1935.
ALSO KNOWN AS: Snake Island Bridge, Ponte Almirante Alexandrino de Alencar, Ponte Pensil Alexandrino de Alencar.
DESIGNED BY: Unidentified.
ENGINEERED BY: Unknown German Consortium.
OPENED: 1915.
TYPE: Suspension bridge, with overshooting main beam, and several design features in common with the 1905 Widnes–Runcorn Bridge.
DISTANCE BETWEEN TOWERS: 168 metres.
HEIGHT OF TOWERS: 70 metres.
CLEARANCE ABOVE HIGH WATER LEVEL: 50 metres.
MOTIVE POWER: Direct DC electric drive.
GONDOLA SIZE: Unknown.
GONDOLA CAPACITY: Quoted as 400 standing passengers.

1938: MAARSSERBURG TRANSBORDEUR

In the mid-1930s a decision was made to replace two swing bridges over the Merwedekanaal – now known as the Amsterdam-Rhine canal – near Maarssenbroek with a high level steel bridge. For farmers with horse-drawn carts, that would have meant a lengthy and exhausting detour to the approach ramps to the new bridge in order to get their produce to market. So as an integral part of the new bridge's design, a small gondola was slung beneath the roadway suspended from a rigid frame and operated 'on demand' by the bridge-master. The transporter was only operational for a few months before the outbreak of the Second World War, and was briefly taken out of service when the bridge was damaged during hostilities in 1945. After the war ended, the European Economic Recovery Plan – more usually referred to as the Marshall Plan – poured billions of dollars of US aid and thousands of American tractors into Western Europe. Announced in 1946, the aid plan ran from 1947 until 1951, and farmers quickly abandoned their horse-drawn carts in favour of tractors and, rather than wait for the transporter, took to the high level bridge. The transporter mechanism, largely unused since the end of the war, was eventually removed in 1959 during a refurbishment of the bridge. The road bridge was further refurbished in 2006 and remains in use today.

Illustration: Anon, from Wikipedia Commons.

LOCATION: Over the Merwedekanaal (now Amsterdam-Rhine canal) in Stichtse Vecht, linking Maarssen with Maarssenbroek, Netherlands.
STATUS: Transporter mechanism dismantled 1959.
ALSO KNOWN AS: Maarsserbrug Hoge Brug (High Bridge).
DESIGNED BY: Unknown.
ENGINEERED BY: Unknown.
OPENED: 1938.
TYPE: Simple rigid-cradle transporter mechanism slung beneath road bridge.
DISTANCE BETWEEN TOWERS: 88 metres.
HEIGHT OF TOWERS: n/a.
CLEARANCE ABOVE HIGH WATER LEVEL: 25 metres.
MOTIVE POWER: Electric winch drive operated by bridge-master.
GONDOLA SIZE: Unspecified.
GONDOLA CAPACITY: Unspecified, but believed to have been able to carry a single horse and cart plus pedestrians.

1873: TEES TRANSPORTER BRIDGE (1)

Charles Smith's design for a transporter bridge – he called it a 'Bridge Ferry' – across the Tees was, perhaps, the first such proposal to be fully articulated. He had designed a workable solution to the problem of moving volumes of traffic across the river, but one which was simply ahead of its time. According to his projections, the travelling frame would run beneath the main girder on two sets of rails set 12.2 metres apart, the frame itself having a width of 14.32 metres and even with a full load on the gondola mid-stream, the girder would have a deflection of less than 12cm. Despite giving precise details of the bridge superstructure itself, the report of the proposed bridge in the journal *Engineering* offered no clue as to the dimensions or carrying capacity of the gondola – which was described as the 'suspended lower platform' – although it did suggest it would be fitted with rails to carry railway wagons as well as pedestrians and road vehicles. With a speed of 5 miles per hour, it was estimated that the bridge should be able to transport a total of 1,380,000 tons of pedestrians,

vehicles and goods per annum. Assuming that the gondola would make six crossings per hour, six days per week, each crossing would involve a passenger/freight weight of around 64 tons. His span was slightly wider than the bridge which was eventually built, with slightly less clearance above high water level.

Illustration: Published in *Engineering* 25 July 1873.

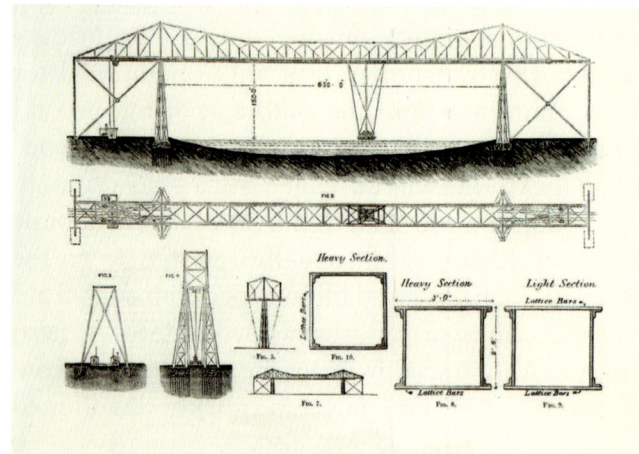

PROPOSED LOCATION: Spanning the River Tees between Middlebrough and Port Clarence, UK.
STATUS: Proposed 1873, never built.
DESIGNED BY: Charles Smith.
ENGINEERED BY: Never built.
TYPE: Cantilevered rigid truss bridge.
DISTANCE BETWEEN TOWERS: 198 metres.
HEIGHT OF TOWERS: 45.7 metres.
CLEARANCE ABOVE HIGH WATER LEVEL: 45.7 metres.
MOTIVE POWER: Steam-driven cable winches.
GONDOLA SIZE: Not specified, but calculated from Smith's drawings at approximately 10m x 8m.
GONDOLA CAPACITY: Based on Smith's estimated annual capacity for the bridge, gondola capacity 64 tons. Total estimated moving weight calculated at around 72 tons.
PROPOSAL DESCRIBED IN: *Engineering*, 25 July 1873.

1878: THAMES MOVABLE BRIDGE

Mills and Twyman's concept for a transporter bridge across the Thames in London owed much in its proposed structure to N.N. Houghton's 1852 design for what he called his 'Aerial Railway bridge' which was described and illustrated in *Scientific American* – the transporter gondola traversing an open central span through which shipping could pass. The maximum clearance under that central span however, at just 21 metres (4 metres lower than the Widnes–Runcorn bridge) would have limited the size of tall-masted vessels which could have passed beneath it. Mills and Twyman did not specify a proposed location for the bridge, but presumably assumed it would be constructed up river from the docks used by the tallest of tall ships.

Illustration: Published in *Engineering* 29 March 1878

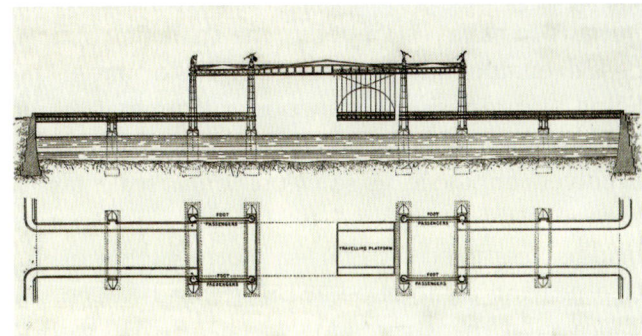

PROPOSED LOCATION: Crossing the River Thames at an unspecified London location, UK.
STATUS: Proposed 1878, never built.
DESIGNED BY: L. Mills and A. Twyman.
ENGINEERED BY: Never built.
TYPE: Steel truss.
DISTANCE BETWEEN TOWERS: 61 metres.
HEIGHT OF TOWERS: 24.4 metres.
CLEARANCE ABOVE HIGH WATER LEVEL: 21 metres.
MOTIVE POWER: Hydraulic drive housed within main beam.
GONDOLA SIZE: Unspecified.
GONDOLA CAPACITY: Estimated by Mills and Twyman as 10 vehicles measuring 6 metres in length or 400 passengers.
PROPOSAL DESCRIBED IN: *Engineering*, 29 March 1878.

1897: TANCARVILLE PONT A TRANSBORDEUR

The bridge Arnodin proposed to construct at Tancarville was intended, in addition to carts and passengers, to carry a tramway, or light railway across the River Seine – in this case *en route* between le Havre and Pont-Audemer. He seems initially to have responded to the challenge of how best to enable railway passage across or under the Seine by proposing a transporter bridge at either Tancarville *or* at Quillebeuf (*qv*). Either would have involved engineering solutions on a scale far in excess of those required for the only transporter bridge completed thus far – at Portugalete – or the two nearing completion at Bizerte and Rouen. At Tancarville, the River Seine was nearly 490 metres wide, compared with 160 metres for the River Nervion at Portugalete. Arnodin's solution was to design a bridge where the towers were sited well out into the river, leaving a clear channel of around 300 metres – still not far short of twice the largest span he had built thus far. The design was essentially the same hybrid of suspension bridge and cable stays used at Portugalete, Bizerte and Rouen, but with the increased loading to which the beam would be subjected being countered by multiple cables tethered on to the approach ramps between the shores and piers, as well as cables embedded in concrete anchors well back from the river. Given the topography of the landscape along the Ribble estuary, it is highly likely that the proposed Ribble Conveyor Bridge (*qv*), put forward the following year, would have required a similar design solution.

Illustration from a supplement to the newspaper *Le Travailleur Normand.*

PROPOSED LOCATION: Across the Seine at Tancarville, Seine-Maritime, Normandy, France.
STATUS: Never built.
DESIGNED BY: Ferdinand Arnodin.
ENGINEERED BY: Never built.
TYPE: Hybrid cable-stayed/suspension bridge.
DISTANCE BETWEEN TOWERS: Estimated at 300 metres.
HEIGHT OF TOWERS: Estimated at 90 metres.
CLEARANCE ABOVE HIGH WATER LEVEL: Estimated at 60 metres.
MOTIVE POWER: Unspecified.
GONDOLA SIZE: Unknown, but to accommodate a single tramcar and other vehicles it would have had to have been at least 15 metres by 8 metres.
GONDOLA CAPACITY: Unspecified, but intended for passengers, vehicles and tramcars.
PROPOSAL DESCRIBED IN: Arnodin, Ferdinand, *Traversée de la Seine Maritime sur Transbordeur à la Pointe de Tancarville,* Orléans 1897, Imprimerie Paul Pigelet; *Le Travailleur Normand,* 27 May 1900 – illustrated in a supplement to the newspaper.

1898: RIBBLE CONVEYOR BRIDGE

The first published mention of the proposed Ribble 'Conveyor Bridge' appeared in the *London Gazette* on 25 November 1898 as part of the advance notification of an intended Bill to be laid before the 1899 Parliamentary session. The bridge was to be part of the tram system linking the two Lancashire resorts of Southport and Lytham, proposed by the newly formed Southport and Lytham Tramroad Company, later known as the Southport District Tramroad Company. Costs were estimated at £183,500. The enabling Act was passed in 1900. A second version – *The Southport and Lytham Tramroad Act 1904* – was also passed, but the project was abandoned in 1909. The company was not dissolved until October 1936.

Had it been built, the central span of the transporter bridge would have exceeded 230 metres to clear the navigable channel giving access to Preston Docks, making it the second longest transporter in the UK, after the Widnes–Runcorn bridge. According to the 1899 proposal, the bridge would be built 'for the purposes of conveying tramcars, carriages, trucks, wagons, passengers, animals, goods and merchandise across the trained channel of the River Ribble'. Either side of the bridge, there would have had to be extensive construction work to enable the tramway and roadway to cross the tidal sands and marshland on both sides of the canalised section of the river. No drawings of the proposed bridge have yet been located.

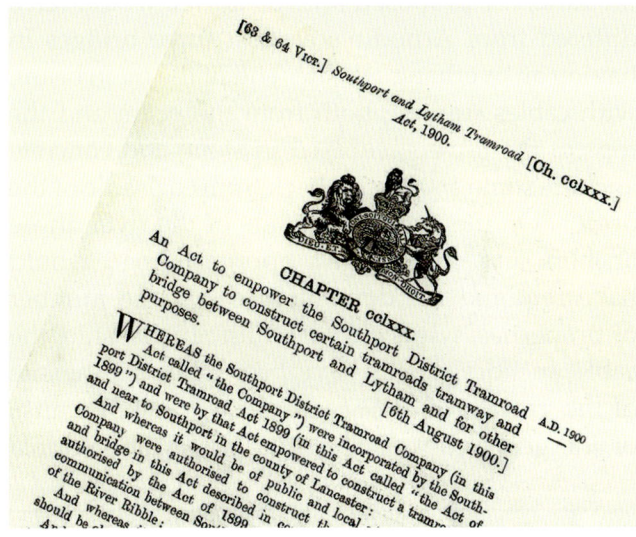

PROPOSED LOCATION: Estuary between North Meols and Lytham, Lancashire, England.

STATUS: Initial proposed 1898, modified 1899, never built.

DESIGNED BY: John T. Wood

ENGINEERED BY: Never built.

TYPE: Suspension bridge with cradled overshooting beam.

DISTANCE BETWEEN TOWERS: 299 metres.

HEIGHT OF TOWERS: Unspecified, but given the required span, the final height would probably have had to be in excess of 80 metres.

CLEARANCE ABOVE HIGH WATER LEVEL: Originally suggested as 31 metres but, to placate the Preston Port Authorities, that was amended to 46 metres and then to 61 metres.

MOTIVE POWER: Described only as 'Suitable Machinery'.

GONDOLA SIZE: Approximately 18 metres x 10 metres.

GONDOLA CAPACITY: Tramcars, road vehicles, passengers, total capacity 25 tons.

PROPOSAL DESCRIBED IN: *The London Gazette*, 25 November 1898; *The London Gazette*, 24 November 1899; *Southport and Lytham Tramroad Act*, 6 August 1900; *Hansard* on fifteen occasions between 19 June 1899 and 11 July 1902, then in 1904, 1906 and 1909; *Modern Tramway Magazine* July 1969, Bond. W. 'Crossing the Ribble'.

1900: SHIELDS TRANSPORTER BRIDGE

The proposed bridge linking South Shields with Tynemouth was put forward by Charles H. Gadsby working with Ferdinand Arnodin so, although relatively little contemporary discussion about the bridge is available, most of what can be gleaned comes from the *Book of Reference* assembled in support of a Private Bill in Parliament. Like the Rocherfort–Martrou bridge nearing completion at the time, the proposed Shields bridge would have differed from Arnodin's earliest three bridges in having an overhang at either end of the main beam, with cables running both from the beam and the tops of the towers, encased in stone and concrete anchors some 160 metres back from each side of the bridge. On the Tynemouth side of the river, those anchors and the bridge's approach road would have required the demolition of a large number of properties, while on the South Shields side, the cable anchors would have been built on the site of the graveyard of St Stephen's Church. Because of the geography of the river bank, the gondola would have crossed the river 19 metres above high water level. Uniquely for a transporter bridge with which Arnodin was involved, the actual open span would have been reduced by having fixed bridge approaches either side of the river – longer on the South Shields side. In November 1900 many of the details had yet to be worked out, the company keeping open the options of powering the bridge 'by means of steam, gas, electricity, or such other mechanical power as they may think fit'.

Illustration: Bridge plan, Durham Record Office.

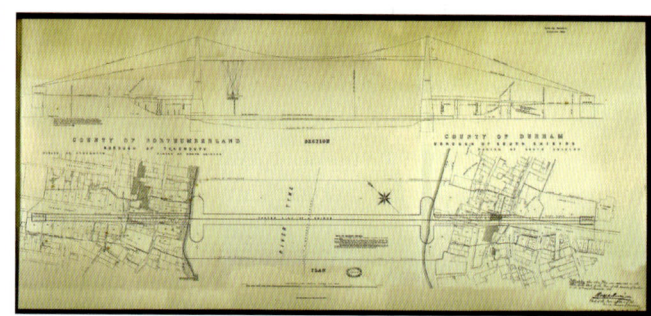

PROPOSED LOCATION: Linking South Shields and Tynemouth, Tyne & Wear, UK.
STATUS: Proposed 1900, never built.
DESIGNED BY: Ferdinand Arnodin & Charles H. Gadsby.
ENGINEERED BY: Never built.
TYPE: Suspension bridge, with overshooting main beam.
DISTANCE BETWEEN TOWERS: 255 metres.
HEIGHT OF TOWERS: 97.5 metres.
CLEARANCE ABOVE HIGH WATER LEVEL: 64 metres.
MOTIVE POWER: Unspecified.
GONDOLA SIZE: Unknown.
GONDOLA CAPACITY: Unspecified.
PROPOSAL DESCRIBED IN: *The London Gazette*, 23 November 1900, *Book of Reference*, now in Durham Record Office.

1902: BORDEAUX TRANSBORDEUR-RICHELIEU

Had it ever been built, the Transbordeur-Richelieu would have been a transporter bridge like no other. For it, Arnodin proposed completely re-inventing the concept of the 'pont transbordeur'. Gone would be the tried-and-tested tall towers and a suspended or cantilevered stiffening beam, and in its place would be the largest single arch bridge yet built. The pair of arches would rise 100 metres above high water level and span a total of 400 metres. The design for each arch would involve three separate hinged sections, while the stiffening beams would be anchored to two pairs of stone towers either side of the river. These would be cross-braced, and below them, the stiffening beams would carry two sets of rails from which would be suspended two gondolas allowing them to pass mid-stream and thus doubling the hourly capacity of the bridge. The design would have created a bridge of immense rigidity without the need for tensioned cables and the huge anchorages required by all of Arnodin's other 'transbordeurs'. From an engineering point of view, it would likely have been a much more expensive bridge to build, and that may be the reason for its abandonment. It has been suggested that the original intention was for the Transbordeur-Medoc to be built to a similar design, but when worked started on that bridge in 1910, it was to a 'conventional' suspension design.

Illustration: from Tyrrell *Transporter Bridges* 1912.

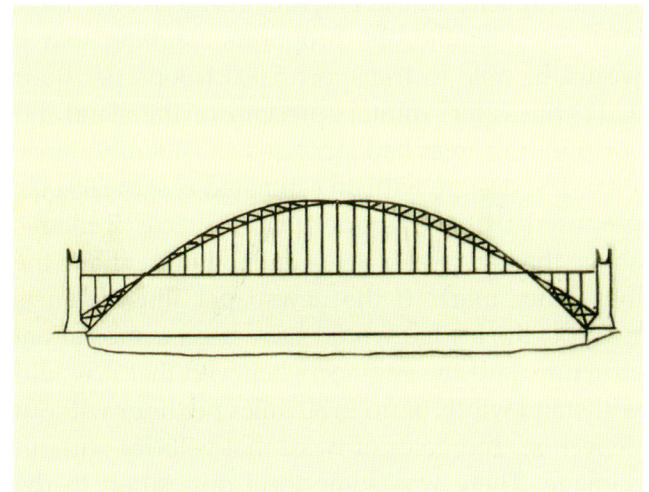

INTENDED LOCATION: Across the Garonne between the Quai de Bacalan near the Place Richelieu and the Quai des Queyries, Bordeaux, Gironde, France.
STATUS: Proposed 1902, never built.
DESIGNED BY: Ferdinand Arnodin.
ENGINEERED BY: Never built.
TYPE: Steel hinged arch bridge.
DISTANCE BETWEEN TOWERS: 400 metres.
HEIGHT OF SUPPORTING ARCH: 99 metres.
CLEARANCE ABOVE HIGH WATER LEVEL: 48 metres.
MOTIVE POWER: not specified.
GONDOLA SIZE: not specified.
GONDOLA CAPACITY: not specified.
PROPOSAL DESCRIBED IN: Henry Grattan Tyrrell *Transporter Bridges,* University of Toronto Engineering Society 1912; *Avant-Projects de Convention de Retrocession du Transbordeur-Richelieu et du Transbordeur-Medoc,* Comité de Patronage des Ponts a Transbordeur de Bordeaux (Systéme Arnodin) December 1902.

1903: HAYLING AERIAL BRIDGE

A newspaper report about the proposed Hayling Island 'Aerial Bridge' in 1903 claimed that 'Similar bridges have been, or are being constructed over the Usk at Newport (Mon.), across the Mersey at Runcorn, and over the Ribble between Southport and Lytham'. While work was underway at both Newport and Runcorn, the Ribble bridge, would never be built.

The capacity of the Hayling bridge would have been considerable, the proposers claiming that it would be able to transport 5,000 troops per hour across the water – military presence on the island and surrounding areas had increased considerably since 1900. The design, with an estimated cost of £68,000, proposed that the gondola would cross 5 metres above the high water level, and 9 metres above the low water mark. If the newspaper illustration is realistic, the bridge would have been a substantial structure, and the engineers believed that it would withstand winds of up to 80 miles per hour without deflection, and gales of twice that velocity without damage. There was some local opposition to the project published in the local press in June 1903, but the 'Conveyor Bridge' was mentioned again in the *Portsmouth and Hayling Light Railway Order* published in April 1905, and in a slightly revised form in November 1909. The rights granted under the Order were extended in May 1910, but the company was wound up the following year. Like Eady's 1905 proposal to span the entrance to Poole Harbour in Dorset, the bridge was never built.

Illustration: Courtesy of *The News*, Portsmouth.

PROPOSED LOCATION: Across the Langstone Harbour Channel from Portsea Island to Hayling Island, Portsmouth Hampshire, UK.

STATUS: Never built.

DESIGNED BY: George Griffin Eady and Wilberforce Cobbett.

ENGINEERED BY: Never built.

TYPE: Suspension bridge.

DISTANCE BETWEEN TOWERS: 220 metres.

HEIGHT OF TOWERS: 49 metres approx.

CLEARANCE ABOVE HIGH WATER LEVEL: 27.4 metres.

MOTIVE POWER: Unspecified.

GONDOLA SIZE: 15 metres x 10 metres.

GONDOLA CAPACITY: Quoted as one electric [tram]car or three or four carriages, plus an unspecified number of pedestrians. Total carrying load 60 tons.

PROPOSAL DESCRIBED IN: *The Evening News*, Portsmouth, 3 April 1903, *Portsmouth and Hayling Light Railway Order* April 1905.

1903: QUILLEBEUF PONT TRANSBORDEUR

For the massive and innovative pont-transbordeur which he proposed building across the River Seine at Quillebeuf, Arnodin joined up with his son-in-law, the maritime engineer and former naval officer Gaston Leinekugel Le Cocq. Had it been built, the bridge would have been an integral part of a railway line between Quillebeuf and Port-Jérôme-sur-Seine. The project exceeded in ambition his proposed bridges at Bordeaux, initial proposals for which had been published just a year earlier. His suggested design used a hybrid combination of cable-stays and suspension cables. Like the Bordeaux bridges, the clear span would have been 400 metres, but in order to safely carry a railway train across the river, it would have required a much more complex design to sustain the variable loads to which the structure would have been subjected. Instead of two pylons, there would have been four, each heavily cable-stayed, to support the weight of the main beam and its heavy load. The total length of the beam would have been 670 metres, and it was predicted that the 80 metre long gondola would have a capacity of around 225 tons.

Arnodin estimated the total cost of the bridge and its approaches at 12 million francs which was, he claimed, one fifth or one sixth of the cost of either a viaduct or a tunnel. The project was re-appraised several times – the last time as late as 1911 when a second railway line between le Havre and Paris was under consideration. A report then suggested that 'the only practically and promptly feasible project – and therefore the most likely to succeed – is the pont-transbordeur système Arnodin', but it was ultimately considered an impractical solution to the railway's needs.

PROPOSED LOCATION: Crossing the River Seine between Quillebeuf and Port-Jérôme.
STATUS: Initial proposed 1903, re-appraised in 1906 and 1910–11, but never built.
DESIGNED BY: Ferdinand Arnodin and Gaston Leinekugel Le Cocq.
ENGINEERED BY: Never built.
TYPE: Hybrid suspension/cable-stayed bridge with over-reaching beam.
DISTANCE BETWEEN TOWERS: 400 metres between inner pair of towers, 670 metres between outer towers.
HEIGHT OF TOWERS: 107 metres.
CLEARANCE ABOVE HIGH WATER LEVEL: 60 metres.
MOTIVE POWER: Unspecified.
GONDOLA SIZE: Projected length of 80 metres, width unspecified.
GONDOLA CAPACITY: Described as being able to transport 'a small railway convoy' with an estimated maximum weight capacity of 225–250 tons.
PROPOSAL DESCRIBED IN: Arnodin, Ferdinand: *Ligne du Havre à Pont-Audemer – Projet de traversée de la Seine entre Port-Jérome et Quillebeuf au moyen d'un transbordeur porte-train – Mémoire explicatif*; Châteauneuf-sur-Loire, Imprimerie E. Duneau, 1905.

1905: TEES TRANSPORTER BRIDGE (2)

This was the second bridge to be proposed as a cost-effective means of linking Middlesbrough with Port Clarence – the first having been Charles Smith's 1873 design – and although it was never built, it undoubtedly speeded up progress towards the construction of the 1911 transporter bridge. The bridge, which would have crossed the river about 200 metres downstream from the present bridge, would have been a cantilevered bridge with a span of 220 metres and a clearance above high water level of 48.7 metres. The gondola, it was proposed, would traverse the river 1.2 metres above the high water level. On the Middlesbrough shore, the towers would be built onto the existing quayside, while on the Port Clarence shore, a 50 metre long fixed bridge would be built out to meet the north towers. The proposal was drawn up in November 1905 by the Middlesbrough and West Hartlepool Light Railway Company as part of their plan to run a tramway between the two towns, and the outline plans and Book of Reference were deposited with Middlesbrough Council in early 1906. The engineers were to be George Griffin Eady

and Alfred S. Frech, and the electrical engineers for the project were named as Hansard and Watkins. The project was debated and eventually rejected by the Council who declared their 'most strenuous opposition to the scheme of the Promoters as far as the Transporter Bridge is concerned.' Their opposition was, in fact, not to the bridge *per se* but, as the Council operated the ferries, it was to the ownership of the crossing rights on the river.

Illustration: Durham County Archives.

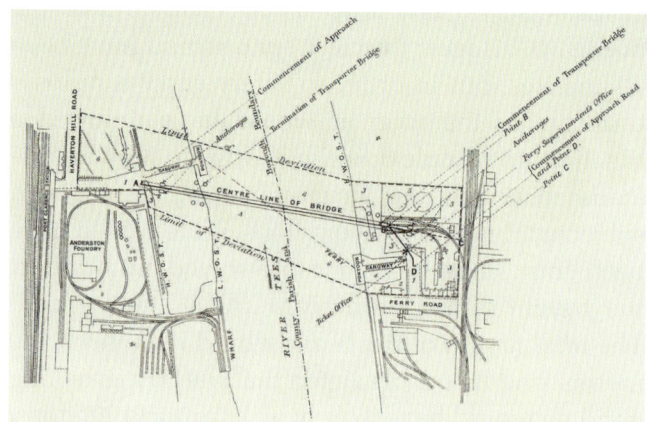

PROPOSED LOCATION: Across the Tees between Middlesbrough and Port Clarence, Middlesbrough, UK.
STATUS: Never built.
DESIGNED BY: George Griffin Eady and Alfred S. Frech.
ENGINEERED BY: Never built.
TYPE: Cantilevered truss with vertical anchorages.
DISTANCE BETWEEN TOWERS: Approximately 220 metres.
HEIGHT OF TOWERS: Approximately 52 metres.
CLEARANCE ABOVE HIGH WATER LEVEL: Approximately 48.5 metres.
MOTIVE POWER: Unspecified.
GONDOLA SIZE: Unspecified.
GONDOLA CAPACITY: Unspecified.
PROPOSAL DESCRIBED IN: *Book of Reference* now in Durham County Archives.

1904–5: POOLE HARBOUR CONVEYOR BRIDGE

George Griffin Eady's enthusiasm for designing transporter bridges was considerable, but, sadly, none of his three proposals was ever built. The Poole Harbour Conveyor Bridge, had it been built, would have had a span of around 400 metres – the biggest ever contemplated in Britain and equal to Arnodin's proposed Bordeaux Transbordeurs. It was proposed as part of the Branksome Park and Swanage Light Railway which called for a:

'Conveyor bridge and approaches of an approximate length of one furlong eight and two-thirds chains commencing in the Parish of Poole and municipal borough of Poole by a junction with Railway (No.3) at its termination crossing the Poole Harbour Entrance Channel and a portion of the Poole Bay of the English Channel in a south-westerly direction and terminating at high water mark of ordinary spring tides at or near the point of land known as South Haven Point in the parish of Studland and rural district of Wareham and Purbeck.'

That would have placed the bridge largely along the same lines as the chain ferry today. Proposed by several local worthies, the tramway and bridge attracted £260,000 of capital guarantees, but the Poole Harbour authorities rejected it in 1906. Griffin and Eady proposed employing one of Arnodin and Palacio's early ideas of using a tethered chain which would be fed through winches mounted underneath the gondola. For such a wide span, this arrangement would have significantly increased the deflection of the main beam as the gondola reached the mid-point of its traverse.

Illustration: Palacio and Arnodin's proposal for a tethered-chain drive system, patented in 1887.

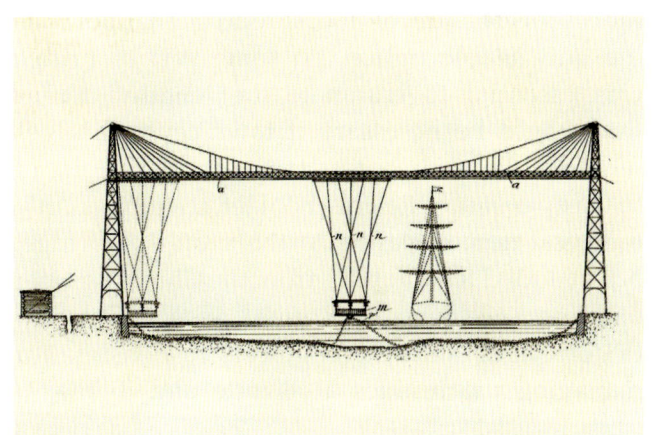

PROPOSED LOCATION: Across Poole Harbour between Sandbanks and South Haven Point, Studland.
STATUS: Proposed 1905, never built.
DESIGNED BY: George Griffin Eady and Alfred S. Frech.
ENGINEERED BY: Never built.
TYPE: Uncradled suspension bridge.
DISTANCE BETWEEN TOWERS: approximately 400 metres.
HEIGHT OF TOWERS: Unknown, but given the span, a height of around 90 metres would have been required.
CLEARANCE ABOVE HIGH WATER LEVEL: Unknown.
MOTIVE POWER: Electric motors in the gondola winding a tethered chain.
GONDOLA SIZE: Unknown but as, in addition to pedestrians, it was to be fitted with 1.07 metres-gauge rails to carry a single tramcar, of average length 8.7 metres, it would have had to have been around 12 metres x 8 metres.
GONDOLA CAPACITY: Single tramcar and unspecified number of foot passengers. A single-decked tramcar of the period had an unloaded weight of around 10 tons.
PROPOSAL DESCRIBED IN: *Branksome Park and Swanage Light Railway Book of Reference*, Dorset History Centre.

1906: OSTENDE PONT A TRANSBORDEUR

The illustration usually associated with the proposed 'Pont-Transbordeur à Ostende' was probably fanciful. While there was clearly interest in the idea of building a transporter bridge across the entrance channel which separated Ostend town centre from 'le Quartier de Phare', the castellated Gothic structure on the cover of the booklet *Ostende et Ses Merveilles* was probably simply for illustrative purposes. Had a transporter bridge ever been built, it would most likely have been modelled on Arnodin's designs for the bridges at Rouen, Marseille or Newport.

Ostend needed to expand, and expanding on the opposite side of the shipping channel was the only viable option, so some sort of bridge was essential to eliminate the lengthy detour that would be necessary if town-centre bridges remained the only access to the proposed site of that expansion. Gielen's booklet contained little technical detail relating to the proposed bridge, but included testimonials from the Chief Engineer of the port of Rouen, the British Consul in Ostend – Roger Gage – and the engineers responsible for managing the bridges at Bizerte and Bilbao. It did, however, stipulate that, if built, the bridge's beam would have to be 54 metres above high water level. Access to the proposed bridge was to have been at the east end of Rue Longue – now known as Langestraat – crossing the channel in a north-easterly direction. Construction, it was estimated, would take one year, with the bridge open to traffic within two years of the project getting official approval.

Illustration: From *Ostende et ses Merveilles*, 1906, courtesy of Ghent University Library.

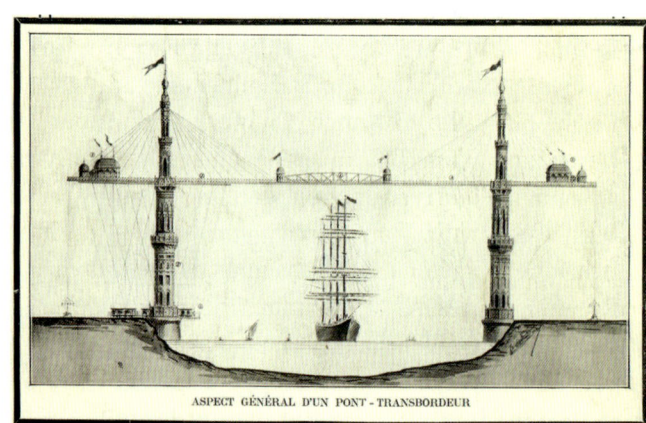

ASPECT GÉNÉRAL D'UN PONT - TRANSBORDEUR

PROPOSED LOCATION: Across the entrance channel to the port of Ostend, Belgium.
STATUS: Proposed 1906, never built.
DESIGNED BY: Unknown.
ENGINEERED BY: Never built.
TYPE: Cable-stayed bridge with cantilevered central section.
DISTANCE BETWEEN TOWERS: 61 metres.
HEIGHT OF TOWERS: approximately 83 metres.
CLEARANCE ABOVE HIGH WATER LEVEL: 54 metres.
MOTIVE POWER: Electric winches in elevated winch-house on the overhanging stiffening beam.
GONDOLA SIZE: 8 metres x 30 metres.
GONDOLA CAPACITY: Proposed as three tramcars, an unspecified number of vehicles and up to 240 passengers.
PROPOSAL DESCRIBED IN: *Ostende et ses Merveilles: le Pont Transbordeur*, Des Presses de Alex. Gielen, Bruxelles, 1906.

1910: BORDEAUX TRANSBORDEUR-MEDOC

The idea of building massive transporter bridges across the Garonne at Bordeaux was first discussed as early as 1893 – at around the time of the completion of Palacio and Arnodin's first bridge – but planning did not get underway until 1903, by which time Arnodin was proposing to build two bridges, one of which, the Transbordeur-Richelieu, would have two gondolas able to pass mid-stream. Construction work on the Transbordeur-Médoc – originally intended to be the second of Bordeaux's 'transbordeurs' – began in September 1910 when the President of the Republic laid the foundation stone. Arnodin's original plans for the Transbordeur-Medoc had suggested it could have one or two gondolas, but the multi-gondola idea seems to have been abandoned long before construction started. By the end of 1913 the towers were complete but the First World War intervened and the project was suspended. Work never restarted after hostilities ceased and what would have been one of Arnodin's most ambitious transporter became nothing more than a white elephant, its tall and elegant towers remaining as Bordeaux landmarks until demolished by the Germans in 1942 to avoid them being used to help guide Allied bombers. It turned out to be the last transporter bridge project to bear Arnodin's name, although he would act as a consultant on later bridges. The stone bases of the towers on the Quai des Queyries, and the concrete and stone cable anchorages sited well behind them still survive.

Illustration: From a postcard looking across the river from the Quai des Queyries, published c.1913.

LOCATION: Across the Garonne between the Quai des Queyries on the east and Coeurs du Médoc on the west, Bordeaux, France.
STATUS: Towers only erected. Demolished, 1942.
ALSO KNOWN AS: Bordeaux Pont Transbordeur.
DESIGNED BY: Ferdinand Arnodin.
ENGINEERED BY: Ferdinand Arnodin.
OPENED: Never completed.
TYPE: Suspension – the lightweight steel lattice beam was to have been supported by vertical cables suspended from a pair of semi-rigid double arcs of latticed steelwork with an articulated coupling mid-span.
DISTANCE BETWEEN TOWERS: 400 metres.
HEIGHT OF TOWERS: 92.5 metres.
CLEARANCE ABOVE HIGH WATER LEVEL: 45 metres planned.
MOTIVE POWER: Intended to have been DC electric motors.
GONDOLA SIZE: Proposed at 10 metres x 13 metres.
GONDOLA CAPACITY: 150 tons intended total moving load.
PROPOSAL DESCRIBED IN: Avant-Projects de Convention de Retrocession du Transbordeur-Richlieu et du Transbordeur-Medoc, Comité de Patronage des Ponts a Transbordeur de Bordeaux (Systéme Arnodin) December 1902.

1928: DUBLIN TRANSPORTER BRIDGE

The proposed transporter bridge across the River Liffey in Dublin was the brainchild of Joseph Mallagh, Chief Engineer of the Dublin Port & Docks Board and Pierce Purcell, Consultant Engineer. Legal powers to build it were enshrined in the Dublin Port and Docks (Bridges) Act, 1929. Mallagh prepared the Parliamentary Bill for the replacement of the swivel Butt Bridge over the River Liffey with a fixed span and also the proposed transporter bridge further downstream. The new Butt Bridge was completed in 1932 but the transporter was not built. Had it been, it would have crossed the river at the same point as today's Samuel Beckett Bridge. It would have had two gondolas able to pass mid-stream – an idea first proposed by Arnodin in 1901 for his Transbordeur-Richelieu in Bordeax which, like the Dublin bridge, was never built. The proposal was revisited several times between 1929 and December 1948, and its costs (estimated at £167,000) debated in the Irish Parliament. Its ultimate abandonment followed shortly thereafter. The project was, effectively, dead long before then. The Docks Board had held a conference in 1936 with the Dublin Corporation and the Dublin County Council at which it was decided that an alternative type of bridge to that proposed in the Act should be considered – an understandable decision in view of the huge increase in motor traffic and major innovations in bridge construction – but a final decision was not taken until 1949, by which time a transporter bridge could not have coped with the traffic volume.

Illustration: courtesy Dublin Port Company Archives.

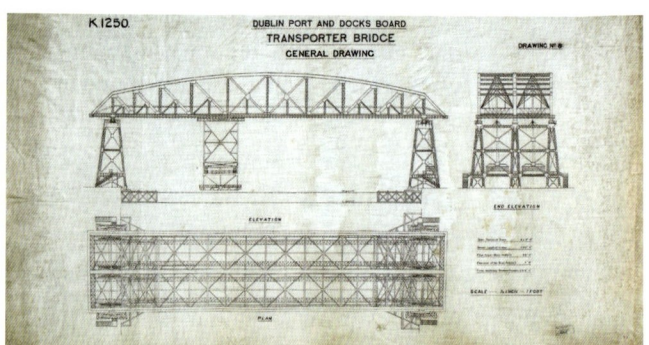

PROPOSED LOCATION: Across the River Liffey, at Guild Street, Dublin, Eire.
STATUS: Never built.
DESIGNED BY: Joseph Mallagh and Pierce Purcell.
ENGINEERED BY: Never built, but Purcell was to have been the Consultant Engineer.
TYPE: Twin self-supporting lattice trusses supported on trapezoid steel towers.
DISTANCE BETWEEN TOWERS: 120 metres.
HEIGHT OF TOWERS: Estimated at approximately 30 metres.
CLEARANCE ABOVE HIGH WATER LEVEL: 30 metres approximately, based on the model's dimensions.
MOTIVE POWER: Simply described as 'hydraulic or electric machinery' in the Parliamentary Bill.
GONDOLA SIZE: Two gondolas of unknown specification operating side by side.
GONDOLA CAPACITY: 16 vehicles and 700 passengers each, crossing the river every three minutes and passing mid-stream.
PROPOSAL DESCRIBED IN: Dublin Port and Docks (Bridges) Act, 1929 and in reports of discussions, amendments and extensions to that Act in the proceedings of the Irish Parliament until 1948. A model of the bridge was illustrated in the Irish Independent on 19 September 1928 and committee proceedings reported in The Evening Herald, Dublin, 20 November 1928.

1998: LONDON ROYAL VICTORIA DOCK BRIDGE

Sixty years after the completion of the last transporter bridge – over the Merwedekanaal near Maarssenbroek in Holland – it seemed the world was about to get a new transporter, albeit for pedestrians and cyclists only. The Royal Victoria Dock Bridge, which crosses the largest of the three Royal Docks in London's rapidly redeveloping Docklands area, was originally conceived as a high-level pedestrian bridge, with a lower-level transporter gondola operating below it. The first phase of the construction, completed in 1998, saw the high-level bridge open to the public, but as for the proposed transporter, the only immediately obvious signs of that intention are the gondola docking stations on either side of the water. The bridge uses a Fink Truss – a modern interpretation of a design originated in Germany in the 1860s by Albert Fink and now most widely used to support wide roof spans. The gondola was to be a fully-enclosed glass passenger cabin which would cross the water just below the truss. It would be raised up as it started to cross – giving an 11 metre clearance over the pleasure craft which would use the dock – and dip down again to its lower level as it approached the opposite docking station. The development of the dock as a marina never happened, so the gondola was never installed – although a high level maintenance gondola does currently operate just below the truss.

Illustration: An artist's impression of the Royal Victoria Bridge with the gondola in place © Lifschutz Davidson Sandilands.

LOCATION: Across the Royal Victoria Dock, London, UK.
STATUS: Operational as a lift-accessed high-level footbridge.
DESIGNED BY: Lifschutz Davidson Sandilands.
ENGINEERED BY: Techniker (main contractor) and Allott and Lomax (services contractors).
OPENED: Bridge opened 1998, transporter never installed.
TYPE: Steel Fink Truss cable-stayed bridge. The design of the truss was intended to reflect the shapes of the sailing craft which, it was planned, would be moored below.
DISTANCE BETWEEN TOWERS: 130 metres.
HEIGHT OF TOWERS: 45 metres.
CLEARANCE ABOVE HIGH WATER LEVEL: 15 metres. Gondola clearance planned at 11 metres above water level.
MOTIVE POWER: Intended to have been cable-driven.
GONDOLA SIZE: Proposed at 8 metres by 3.2 metres.
GONDOLA CAPACITY: Up to 45 people, or a combination of people and bicycles.
PROPOSAL DESCRIBED IN: Competition documents submitted by Lifschutz Davidson Sandilands.

INDEX